SOLID STATE CHEMISTRY

Solid State Chemistry

Synthesis, Structure, and Properties
of Selected Oxides and Sulfides

Aaron Wold
and
Kirby Dwight

CHAPMAN & HALL, Inc.
New York

First published in 1993 by
Chapman and Hall
29 West 35th Street
New York, NY 10001-2299

Published in Great Britain by
Chapman and Hall
2-6 Boundary Row
London SE1 8HN

Cover Illustration: A portion of the structure of Rh_2O_3 (III) as viewed along the orthorhombic x-axis.

Library of Congress Cataloging in Publication Data

Wold, Aaron, 1927–
    Solid state chemistery : synthesis, structure, and properties of
selected oxides and sulfides / Aaron Wold and Kirby Dwight.
        p.   cm.
    Includes bibliographical references and index.
    ISBN 0-412-03611-8. — ISBN 0-412-03621-5 (pbk.)
    1. Solid state chemistry.   I. Dwight, Kirby, 1928–
II. Title.
QD478.W65   1993
541'.0421—dc20   92-35607
                                                        CIP

British Library of Congress Cataloguing in Publication Data also available

Contents

Preface

The subject matter of solid state chemistry lies within the spheres of both physical and inorganic chemistry. In addition, there is a large overlap with solid state physics and materials engineering. However, solid state chemistry has still to be recognized by the general body of chemists as a legitimate subfield of chemistry. The discipline is not even well defined as to content and has many facets that make writing a textbook a formidable task.

The early studies carried out in the United States by Roland Ward and his co-workers emphasized the synthesis of new materials and the determination of their structure. His work on doped alkaline earth sulfides formed the basis for the development of infrared phosphors and his pioneering studies on oxides were important in understanding the structural features of both the perovskite oxides as well as the magnetoplumbites. In 1945, A. F. Wells published the first edition of *Structural Inorganic Chemistry*. This work attempts to demonstrate that the synthesis, structure, and properties of solids form an important part of inorganic chemistry. Now, after almost 50 years during which many notable advances have been made in solid state chemistry, it is still evident that the synthesis, structure determination, and properties of solids receive little attention in most treatments of inorganic chemistry.

The development of the field since the early studies of Roland Ward (early 1940s) has been rapid. It was shown during this period that most of the previous attempts to write such a book have resulted in either a modified solid state physics text or have been deficient in dealing with the problems of careful chemistry. In writing this book, the authors were forced to make difficult decisions as to which material would be covered for a one-semester course. At the same time, it was realized that a book emphasizing reproducible, well-documented chemistry would appeal to a broader audience. In the choice of the material discussed, emphasis was placed on including compounds that the two authors had worked on since their collaboration began in the mid-1950s. The choice was limited primarily to transition metal oxides and sulfides, although with a few exceptions, e.g., zinc sulfide, tin sulfides, and diamond have been included.

The field of solid state chemistry has also undergone considerable change since the 1950s. During the early years, the discovery of magnetic ferrites, semiconductors, superconductors, phosphors, cracking catalysts, etc., stimulated the search for new materials. The relationship between synthesis and structure was considered to be the domain of the chemist, whereas the physicist characterized a few relatively well-defined substances and applied new theoretical models to explain their properties.

Simultaneously, ceramists and metallurgists developed new methods for the processing of materials into useful components. The blurring of disciplines took place as a result of the creation of interdisciplinary groups, which were needed to prepare and characterize more complex materials as well as to understand the relationships among synthesis, structure, and properties. The hope of many chemists was that someone else would discover the merit of their new compounds. The property investigators made measurements wishing that other materials, or better, were available. Bringing these different disciplines together has created "solid state scientists."

This fusion of disciplines has also made it exceedingly difficult to define solid state chemistry, no less write a book that would appeal to a broad audience. The need to make available an introductory textbook that would relate careful chemistry and characterization of solids with a simple model capable of relating properties to composition and structure is a goal desired by solid state chemists, but quite often unattainable. All of the studies discussed in this book have been reproduced by numerous researchers and reflect the research aims and goals of the writers.

It is the hope of the authors that this treatment of solid state chemistry will also be suitable as an introductory textbook for the synthesis and characterization of oxides and sulfides. There are five introductory chapters, which deal with the elements of crystallography, properties of solids, and phase diagrams. The level of the physics is such that any senior chemistry student should be able to follow the discussions of properties with little difficulty. *This tutorial is meant only to provide the reader with the basic language necessary to understand the following sections.* Further treatment of these topics can be found in the references listed after each tutorial chapter.

The tutorial section is followed by a treatment of the synthesis of oxides as both polycrystalline solids and single crystals. There is then a discussion of a number of binary as well as ternary oxides. The various compounds selected are those that one of the authors (Wold) has synthesized in his own laboratories or has confirmed the purity reported by other chemists. The compounds are classified according to structure type starting from simple binary phases to ternary or more complex phases. The theoretical model used to correlate properties and/or structure was developed in the 1960s by J. B. Goodenough and the model is still useful today. A similar treatment follows for a number of the sulfides. This choice of material may, therefore, appear to limit the scope of the book, but it is the authors' desire to discuss the synthetic techniques and properties of materials that are known to be reproducible in the laboratory by the reader if he or she is of a mind to do so.

The problems that appear in this book are meant to test the student's ability to apply the principles discussed in the various sections. They also require specific knowledge that can be acquired only by reading the referenced material.

The authors would like to thank their many students, who have provided them with new insights into synthetic solid state chemistry. They have made it possible to verify the syntheses and understand the properties of many of the oxides and sulfides that are described in this book. We would also like to thank Margery Pustell who designed the cover of the book and assisted in all phases of the preparation of this manuscript.

A.W.

K.D.

PART I
Tutorial

1

Crystal Structure

The nature of the arrangement of atoms or molecules in a material in large measure influences the degree of order and hence its crystallinity. In a well-crystallized compound, the atomic array is periodic. Periodicity is related to the regular repetition of a representative unit of the structure along the directions in the crystal. The periodic atomic arrangement influences all of the properties, e.g., electronic, optical, and mechanical, of the material. To study the relationships between crystal structure and other properties, it is necessary to consider the different ways in which periodic arrays can exist in a crystalline substance.

For a crystal, there are several fundamental ways in which a representative unit or motif can be repeated in space and these will be discussed below. In amorphous materials, the same units are arrayed in a more random fashion. Before giving a detailed treatment of the symmetry operations that can be performed on a structure, it would be useful to relate translation to the arrangement of atoms or molecules in a structure.

A. Translation

If the symbol ↑ is used to symbolize a basic group of atoms or molecules in a crystal, then Fig. 1.1a indicates a translation by an amount t along a direction in space. The combination of the translation shown in Fig. 1.1a with another non-collinear translation t_2, gives rise to the lattice array shown in Fig. 1.1b. The replacement of ↑ in Fig. 1.1b by a point results in a collection of points and is called a two-dimensional plane lattice. It is possible to add a third translation t_3 which would then produce a space lattice in three dimensions. This is shown in Fig. 1.1c. All of the cells represented are identical, so that we can choose any one, e.g., the heavily outlined one, to be a unit cell. Such a cell (Fig. 1.2) can be described by three vectors a, b, and c which are drawn from the origin to three corners of the cell. The three vectors that define the cell are the crystallographic axes of the cell and are

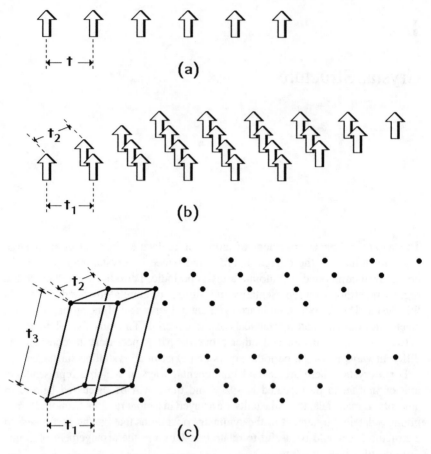

Figure 1.1. Translation of a motif along (a) one direction in space, (b) two non-collinear directions, and (c) a three-dimensional space lattice.

defined in terms of their lengths (a, b, c) and the angles between them (α, β, γ). These lengths and angles are the lattice parameters of the unit cell.

B. Crystal Planes and Directions

Consider in Fig. 1.3 a set of three non-coplanar axes having lengths a, b, c that are cut by the plane ABC thus making the intercepts OA, OB, and OC. If unit lengths are chosen for a, b, and c, then the lengths of the intercepts can be expressed as OA/a, OB/b, and OC/c. The reciprocals of these lengths would be a/OA, b/OB, and c/OC. The small integers h, k, and ℓ can be defined in terms of these reciprocal lengths, i.e. $h=a/OA$, $k=b/OB$, and $\ell=c/OC$. This defines

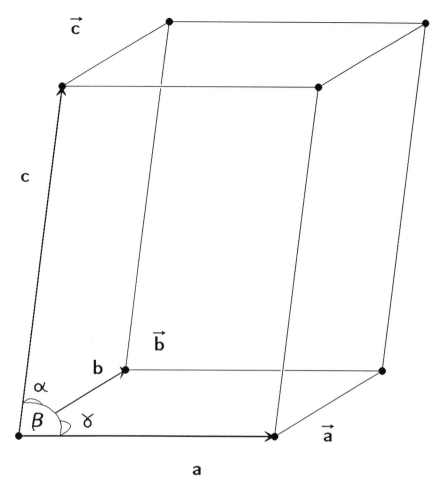

Figure 1.2. Relationship between the angles and edges of a unit cell.

the law of rational intercepts. Haüy and Miller in 1839 proposed the use of $(hk\ell)$ values as indices used to define the crystal faces.

If a face is parallel to an axis, the intercept is at ∞ and the Miller index becomes $1/\infty$ or 0. Figure 1.4 illustrates the Miller indices for several crystal planes, and planes of atoms parallel to those shown have the same indices. If an axis (e.g., x axis) is cut by a plane on the negative side of the origin, the corresponding index will be negative and a minus sign is placed above the appropriate index, e.g. $(\bar{h}k\ell)$. Whereas $(hk\ell)$ refers to a single plane or a set of parallel planes, curly brackets { } signify planes of a form, e.g., equivalent planes for cubic symmetry: $\{100\} = (100) + (010) + (001) + (\bar{1}00) + (0\bar{1}0) + (00\bar{1})$.

The indices of a direction are obtained in a different way. Consider a point at

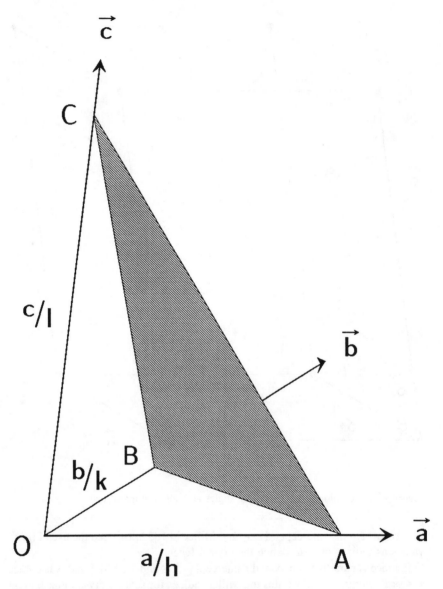

Figure 1.3. Crystal axes intercepted by a crystal plane.

the origin of coordinates that must be moved in a given direction by means of motion parallel to the three crystal axes. Suppose this can be accomplished by going along the *x* axis a distance *u***a**, along the *y* axis *v***b** and along the *z* axis *w***c**. The integers *u, v, w* are the indices of the direction and are written [*uvw*]. It should be noted that reciprocals are not used in calculating the indices of a

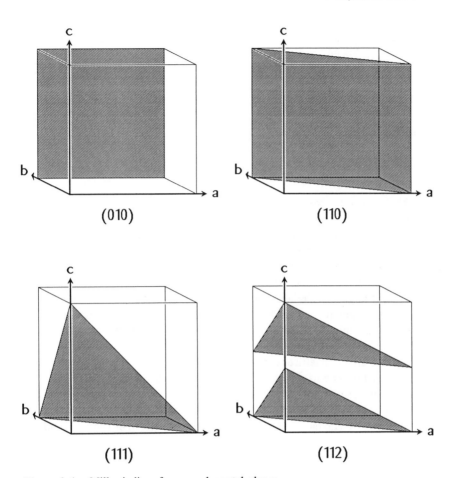

Figure 1.4. Miller indices for several crystal planes.

direction. A full set of equivalent directions is indicated by angle brackets $<uvw>$.

C. Bravais or Space Lattices

In a point (or Bravais lattice), the repetition of primitive translations will lead from one lattice point to any other. The important characteristic of a space lattice is that every point has identical surroundings. There are only 14 ways in which points can be arranged in space so that each point has identical surroundings. Although there are a great many crystal structures, each of these consists of some fundamental pattern repeated at each point of a three-dimensional space lattice. The space lattices are limited in number, whereas there are many crystal struc-

Table 1.1. Crystal Systems

System	Axes	Interaxial angles
Cubic	$a = b = c$	$\alpha = \beta = \gamma = 90°$
Tetragonal	$a = b \neq c$	$\alpha = \beta = \gamma = 90°$
Orthorhombic	$a \neq b \neq c$	$\alpha = \beta = \gamma = 90°$
Monoclinic	$a \neq b \neq c$	$\alpha = \gamma = 90°, \beta$
Rhombohedral	$a = b = c$	$\alpha = \beta = \gamma$
Hexagonal	$a = b \neq c$	$\alpha = \beta = 90°, \gamma = 120°$
Triclinic	$a \neq b \neq c$	α, β, γ

tures. The term lattice has often been incorrectly used as a synonym for structure, a practice that is apt to be confusing.

A given arrangement of points in a space lattice, or of atoms in a structure, is specified by giving their coordinates with respect to a set of axes chosen with the origin at one of the lattice points. Hence, each space lattice has a convenient set of axes, some being of equal length and others unequal. In addition, some axes stand at right angles and others not. There are seven different systems of axes used in crystallography and each possesses certain characteristics as to equality of lengths and angles. The seven crystal systems are given in Table 1.1 along with their axial lengths and angles as given in Fig. 1.2.

Seven different point lattices can be obtained by placing points at the corners of the seven crystal systems. However, one important characteristic of a lattice is that each point has identical surroundings so that other arrangements of points are possible. The French crystallographer, Bravais, demonstrated that there are 14 possible lattices and no more. These Bravais lattices are given in Table 1.2 and are

Table 1.2. Fourteen Bravais Lattices

System	Bravais lattice	Lattice symbol
Cubic	Simple	P
	Body centered	I
	Face centered	F
Tetragonal	Simple	P
	Body centered	I
Orthorhombic	Simple	P
	Body centered	I
	Base centered	C
	Face centered	F
Rhombohedral	Simple	R
Hexagonal	Simple	P
Monoclinic	Simple	P
	Base centered	C
Triclinic	Simple	P

illustrated in Fig. 1.5. The symbols P or R refer to simple or primitive cells that have only one lattice point per cell; non-primitive cells have more than one. The symbols F and I refer to face-centered and body-centered cells, respectively, whereas A, B, and C refer to side-centered cells centered on one pair of opposite faces A, B, or C. The A face is the face defined by the *b* and *c* axes and similarly for B and C. The symbol R is reserved for the rhombohedral system.

As described above, the Bravais lattices determine the repetition of a representative unit or motif in space. In addition, there are several symmetry operations that can be performed on a structure that may bring it into

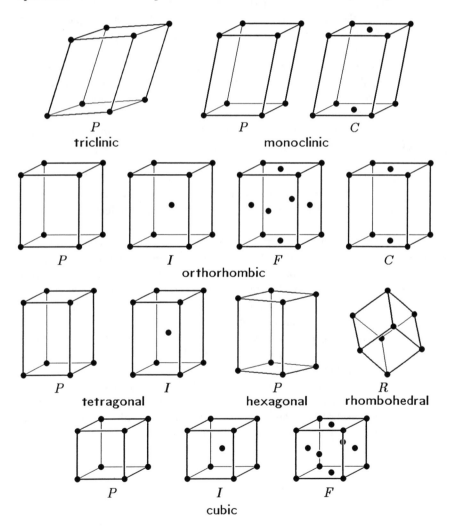

Figure 1.5. The fourteen Bravais lattices.

coincidence with itself. These operations are rotation, reflection, inversion, and rotation–inversion.

D. Symmetry Elements

A structure has n-fold rotational symmetry about an axis if a rotation of $360°/n$ brings it into coincidence with itself. Crystals can show rotational symmetries of 1, 2, 3, 4, and 6. Five-fold symmetry has not been observed since it is not possible to pack together pentagons to form a close-packed layer. The two systems for designating the various symmetry elements are the Hermann–Mauguin system and the Schönflies system. The former is used by crystallographers and the latter by spectroscopists.

The Hermann–Mauguin notation for rotation axes is given by the symbols 2, 3, 4, 6. These symbols imply self-coincidence after rotations through 180°, 120°, 90°, and 60°, respectively. The axes are commonly called diads, triads, tetrads, and hexads. The Schönflies notation designates a proper rotation by the symbol C_n. In this book, the Hermann-Mauguin notation will usually be used and, in the discussion of the other symmetry elements, Schönflies notation will be given in parenthesis.

The reflection through a plane of symmetry or mirror plane is designated by the symbol m (σ) and brings the structure back into coincidence with itself. Hence, a plane of symmetry produces an indistinguishable reflection of all parts of a structure.

A center of symmetry (or inversion center) is identified by the symbol $\bar{1}$, one bar (i) and is what the name implies. If the origin, in Cartesian coordinates, is placed on the center of symmetry, then for the point (xyz) there must be a symmetry related point at ($-x, -y, -z$). In Fig. 1.6, the operation $\bar{1}$ (i) is seen to make equivalent to point P a second point P', which is now on the opposite side of the origin.

An inversion axis combines the operation of simple rotation with inversion through a center of symmetry. This operation is designated by the symbols $\bar{3}$, $\bar{4}$, $\bar{6}$. A 2-fold inversion axis is equivalent to a reflection plane m, and hence this latter symbol is generally used.

It is possible to combine a proper rotation about an axis with a reflection perpendicular to this axis into a single operation of repetition. This operation is called an improper rotation and Fig. 1.7 illustrates a 4-fold rotation combined with a reflection to give a 4-fold rotoreflection axis $\bar{4}$ (S_4). A rotoinversion axis has an equivalent rotoreflection axis: $\bar{2} = \bar{1}$, $\bar{3} = \bar{6}$, $\bar{4} = \bar{4}$, $\bar{6} = \bar{3}$.

E. Screw Axes and Glide Planes

The symmetry elements discussed so far are the elements of point symmetry, which are useful in the study of molecular structure. Finite-sized molecules

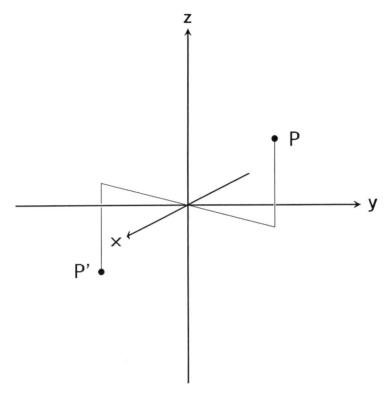

Figure 1.6. Operation of the inversion $\bar{1}$ (or i) on a point P.

possess only point symmetry elements. However, for crystals, it is necessary to add symmetries that include a translation component.

It is possible to combine a rotation and a translation parallel to the rotation axis. Such a combination of symmetry elements is called a screw axis and is designated by the symbols $2_1, 3_1, 3_2, 4_1, 4_2, 4_3, 6_1, 6_2, 6_3, 6_4, 6_5$. The first number indicates the amount of rotation as with simple rotation axes. The subscript divided by the first number indicates the amount of the translation expressed as a fraction of the repeat distance in the direction of the axis. Thus, for axes parallel to the z direction 2_1 has a translation $c/2$, 4_1 has a translation $c/4$, 6_2 has a translation $2c/6 = c/3$, etc.

A glide plane combines a reflection plane with a translation parallel to the plane. The structure is brought into coincidence with itself by reflection across the plane and simultaneous movement along the plane a specified distance. In Fig. 1.8, the point down ↓ is produced from the point up ↑ by the action of glide plane AA. The different kinds of glide planes have translations which are half the axial lengths, half the face diagonals, or one-fourth the face diagonals. The symbols for the representation of glide planes are

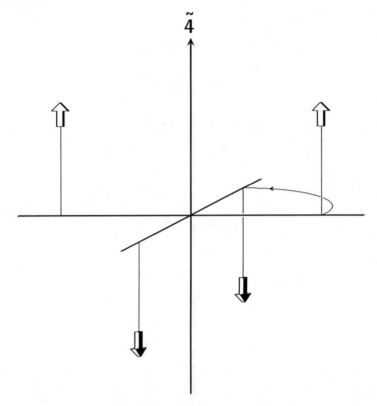

Figure 1.7. Operation of a 4-fold rotoreflection axis $\tilde{4}$ (S_4) on a motif.

a: A glide parallel to the *x* direction with translation *a*/2.
b: A glide parallel to the *y* direction with translation *b*/2.
c: A glide parallel to the *z* direction with translation *c*/2.
n: A diagonal glide with translation ½*a* + ½*b* (or ½*b* + ½*c* or ½*c* + ½*a*).
d: A diamond glide with translation ¼*a* + ¼*b* (or ¼*b* + ¼*c* or ¼*a* + ¼*c*). This glide operation is a feature of the diamond structure.

As far as the external symmetry of crystals is concerned, glide planes cannot be distinguished from reflection planes or screw axes from rotation axes of the same multiplicity.

F. Space Groups

Symmetry of the external faces of a crystal as well as its anisotropic physical properties allow classification into one of 32 crystal classes, which arise from the

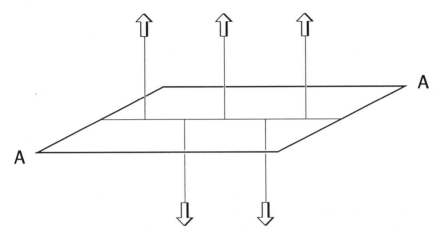

Figure 1.8. Operation of a glide plane on a motif.

seven crystal systems. The symmetry of each class can be described by macroscopic symmetry elements. Equivalent points can also describe each crystal class and are written in the form xyz, $\bar{x}\bar{y}\bar{z}$, etc. The operation of all the symmetry elements of the point group on any one of the equivalent points will produce all the others. However, to describe a group of symmetry elements in space, it is necessary to consider the symmetry elements in three dimensions on a space lattice. This is the basis for classifying crystals into one of 230 space groups. The symmetry elements of a space group, operating on a point located at random in a unit cell of the lattice, will produce a set of equivalent points in an equivalent cell. If an atom is located at one of these equivalent points, identical atoms must be found at each of the other equivalent points.

The symmetry elements of each of the 230 space groups can be found in the *International Tables for Crystallography* (1). A complete treatment of the classification can also be found in this reference. What follows is a brief summary of some X-ray criteria for the classification of crystals and a few examples that describe several space groups.

A special symbolism is used by crystallographers to describe the combination of lattice type and symmetry elements present in a crystal. The shorthand notation first lists lattice type, followed by the principal symmetry elements that are evident either parallel or normal to the unique directions in a unit cell. The symbol for each of the crystal systems shows a minimal symmetry which is characteristic of that system (Table 1.3).

Consider the space group symbol $I4/mcm$. This space group contains only one 4-fold axis and, therefore, as seen from Table 1.3, belongs to the tetragonal system, in which the two crystallographic axes perpendicular to the c axis are

Table 1.3. Minimal Symmetry for the Crystal Systems

System	Minimal symmetry
Triclinic	1 (or $\bar{1}$)
Monoclinic	2 (or m)
Orthorhombic	222 (or $mm2$ or mmm)
Tetragonal	4
Cubic	3 (four 3-fold axes)
Hexagonal	6 (or 3 for trigonal)
Rhombohedral	3

equivalent ($a = b$). The unique directions are the [001] (c axis), [100] (a axis), and [110]; the listing of symmetry elements is given in this sequence. Consequently, the space group symbol is interpreted as follows: The unit cell is body centered (I), the 4-fold rotation along the c axis is perpendicular to a mirror plane (4/m), a glide plane (c) with a translation of ½c is perpendicular to the a axis, and a mirror plane (m) is perpendicular to the [110].

The monoclinic system permits only a single diad and/or a single mirror/glide plane. If another diad or mirror is present, then the crystal system must be orthorhombic. Consider the space group symbol $P2_1/c$. The presence of a single diad (2-fold axis) denotes monoclinic symmetry, as seen from Table 1.3. The only unique direction is the [010] (b axis), and the listed symmetry elements refer to this direction. Thus the symbol is interpreted to mean a primitive monoclinic lattice (P) with a screw diad (2_1) along the b axis and a glide plane /c with translation of ½c perpendicular to it.

For the orthorhombic system, symmetry elements are listed in the sequence [100] (a axis), [010] (b axis), [001] (c axis). The space group symbol $Pbcm$, therefore, indicates a primitive Bravais lattice (P), a glide plane (b) with a translation of ½b perpendicular to the a axis, a glide plane (c) with a translation of ½c perpendicular to the b axis, and a mirror plane (m) perpendicular to the c axis. It is not necessary to explicitly list all the actual symmetry elements. The diad axis, screw diad axes, and centers present are not included in the above symbol because they are not required to define the space group uniquely. However, it is important to observe the significance of the symbols, which refer to the a, b, and c directions in turn.

Finally, the following points should be noted concerning the sequence of operations following the lattice symbol:

1. In all space groups of the unaxial systems, the first symbol (after the lattice symbol) denotes the unique axis—triad, tetrad, or hexad. Since by convention this axis is taken as the z axis, the order of symbols is altered from that used in systems with more than one unique symmetry axis.

2. In the tetragonal system, the second symbol refers to the two equivalent directions parallel to the x and y axes. The third symbol refers to the two equivalent directions bisecting the angles between the x and y axes.

The above were chosen as representative examples of space groups. For a more detailed treatment the reader is referred to the *International Tables for Crystallography*, Vol. A (1).

The application of the symmetry operations of a space group to an arbitrary point generates a set of equivalent points known as the "general" set. When they are applied to a point lying in a position of symmetry (e.g., in a mirror plane, on an axis of rotation, or at a junction of such axes), a smaller "special" set of equivalent points is generated. Different "special" sets are generated from different positions of symmetry. The equivalent points in each space group are listed in tables in such a way as to show the number (multiplicity) of equivalent points belonging to each set and the corresponding position of symmetry. Thus the number of equivalent atoms that can be located at these positions in a crystal can be ascertained.

Consider the orthorhombic space group $P2_12_12$ illustrated in Fig. 1.9. The diad in the c direction is depicted as a full arrow and carries the up arrows ↑ into each other, and similarly the down arrows ↓. The screw diads in the a and b

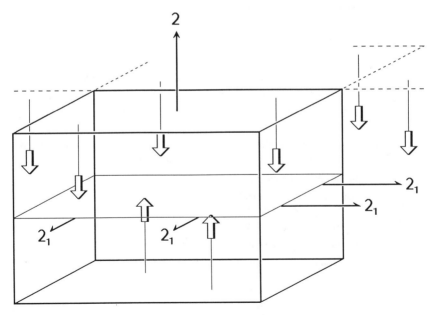

Figure 1.9. Operation of the unique symmetries of the space group $P2_12_12$ on a motif located at a general position. Note that when such an operation carries a motif out of the unit cell, it can always be returned to that cell by lattice translations.

directions are indicated by half arrows and carry up arrows into down arrows and vice versa. These four arrows illustrate the general set of equivalent points for this space group. It should be noted that motifs associated with any point are subject to all the symmetry operations of the space group.

References

1. *International Tables for Crystallography*, Vol. A, Theo. Hahn, ed., D. Reidel, Dordrecht, 1983.

Additional References

L. V. Azároff, *Introduction to Solids*, McGraw Hill, New York, 1960.

C. W. Bunn, *Chemical Crystallography*, 2nd ed. Oxford Press, Oxford, 1961.

B. D. Cullity, *Elements of X-ray Diffraction*, 2nd ed. Addison-Wesley, Reading, MA, 1978.

B. G. Hyde and S. Anderson, *Inorganic Crystal Structures*, John Wiley, New York, 1989.

A. F. Wells, *Structural Inorganic Chemistry*, 5th ed. Oxford Press, Oxford, 1984.

Structure Reports, published yearly by Kluwer Academic Publishers Group. Annual summary of newly determined structures.

Problems

P1.1.   A crystal plane intercepts the three crystallographic axes at the following multiples of the unit distances: 3/2, 2, 1. What are the Miller indices of the plane?

P1.2.   What is the basic lattice of the cesium chloride structure? Why is it not body-centered cubic?

P1.3.   Show that in the orthorhombic system it is possible to have a side-centered and a face-centered lattice but not a two-side centered lattice.

P1.4.   Show that in the tetragonal system $C = P$ and $F = I$.

P1.5.   Show that in a cube-shaped cell
   a.   [111] is perpendicular to (111)
   b.   [100] is perpendicular to (100)

P1.6.   By means of sketches, indicate the screw axes: 6_3 and 6_5.

P1.7.   The spinel structure crystallizes in the space group $Fd3m$. Indicate the significance of each term.

2

Characterization of Solids

The characterization of a solid should describe the features of its composition and structure (including defects) that are significant for the reproduction of the synthesis and for the study of its properties or use. The property measured should reflect directly and unambiguously on the material's composition or structural features. There are many published papers that claim purities of 99.999% (5-9's) or 99.9999% (6-9's) for the starting materials used to synthesize a compound of interest. Such claims are of questionable validity since estimates of purity from such techniques as direct emission spectrographic analysis are in the range of only ≈ 10 ppm for most elements.

The characterization of materials with respect to composition ideally involves identification of all chemical constituents present, as well as the determination of their concentration, electronic state, state of combination, location, and distribution throughout a sample.

A. Characterization of the Major Phase

For elements and for many simple compounds, the matter of major phase identification is trivial. However, in the case of complex compounds, phase identification is prerequisite to detailed characterization. Indeed, problems frequently arise through improper identification of phases present and through the failure to investigate homogeneity.

a. Microscopy

One of the most useful analytical techniques used to characterize the major phase is microscopic examination. Optical and both scanning and transmission electron microscopy are useful for determination of uniformity, inclusions, crystallinity, and particle size. The electron microscope can often be used also for electron diffraction, which is of value in determining crystallinity.

The difficulty in optically identifying or deciphering textures of phases of similar color and reflectivity can often be alleviated by means of staining or etching techniques that can enhance differences between phases. The refractive indices of transparent and opaque solids can be measured by a microscopic technique. All transparent or translucent materials, when immersed in liquids, yield images in the microscope that are bounded by dark shadow outlines or color halos. If the object and surrounding liquid have the same refractive index, and the same color, no line of demarcation will be observable and the object being measured will appear invisible. Hence, given a series of liquids of known refractive indices, the object, or different portions of it, may be immersed in these successively until one is found in which the shadow boundaries are at a minimum, thus determining its refractive index.

The upper limit of magnification with an optical microscope is approximately $2000\times$. Where higher magnifications are desired, an image of the structure can be formed using a beam of electrons rather than light radiation. When electrons are accelerated across large voltages, they can be made to have wavelengths of 0.003 nm. As a consequence of these short wavelengths, the features of atomic clusters can be resolved. The two techniques, which are widely used modes of operation, are transmission and reflection beam modes.

With a transmission electron microscope, the image is formed when an electron beam passes through the specimen. Differences in beam scattering or diffraction produced between various elements of the microstructure form contrasts in the image formed. The preparation of an appropriate sample is essential. Usually, a thin section (or foil) is used, which allows transmission through the specimen of a large fraction of the incident beam. The transmitted beam can be projected onto a fluorescent screen or a photographic film.

In scanning electron microscopy, the surface of a sample is scanned with an electron beam and the reflected, or back-scattered, beam of electrons is collected and displayed at the same scanning rate on a cathode ray tube. The image represents the surface features of the specimen. A limitation is that the surface must be electrically conductive, or else a thin conducting coating must be applied to the surface.

The field of microscopy is undergoing a rapid change and new, more powerful techniques are being developed to permit examination of structure of monolayers of deposited films.

b. Microprobe Analysis

Chemical analysis of a small region within a large sample is called microanalysis. Such analysis can be carried out with an X-ray microprobe or with various forms of the electron microscope, if the latter is equipped with an X-ray spectrometer. The diameter of the analyzed region can be ≈ 1000 Å. The instruments used are collectively known as electron-column instruments. They are all like elaborate X-ray tubes, in which the specimen is the target, and where the electron beam from the filament is focused into a small spot on the sample and when the resulting X-rays are collected and analyzed. The techniques can be applied to small individual particles or scanned across regions of a larger sample in order to determine the extent of compositional variations within the region studied.

c. Density Determination

The density of a solid is easy to determine with relatively high accuracy. It can be obtained either by flotation in liquids of known density or by weighing the solid first in air and again when submerged in water (or other known liquid) and using Archimedes' principle to calculate the density.

The hydrostatic technique was applied by Adams (1) to determine the precise densities of chalcopyrite-deficient phases. The same method has been used widely to determine the density of a variety of materials. An excellent inert density fluid is perfluoro(1-methyldecalin) which can be calibrated with a high purity silicon crystal ($\rho = 1.328$ at $T = 22°C$). Air has a tendency to be adsorbed onto the sample surface or be trapped in the pits and crevices of crystals. This effect causes a decrease in the measured density. To correct this problem, samples must be ground thoroughly and placed in the fluid under vacuum for periods up to 1 hour prior to weighing.

d. X-Ray Diffraction

One of the best methods for the determination of structure of solids is by X-ray diffraction. In a manner similar to that whereby light is diffracted by a grating of suitably spaced lines, the shorter wavelength X-rays can be diffracted. Atoms or ions located in a regular array in a crystalline solid can diffract an X-ray beam and the resulting pattern can be detected by using a photographic film or other device. Diffracted beams from atoms in successive planes can cancel each other unless they are in phase, and to be in phase they must obey the Bragg law

$$n\lambda = 2d \sin \theta$$

where λ is the wavelength of the X-rays, d is the distance between planes, and θ is the angle of incidence of the X-ray beam to the plane. Since the θ values can be measured directly, the values of d can be calculated given the wavelength of the X-rays used. Measurement of the intensities of the diffracted beam in each direction can give a complete structure determination. Each atom in the lattice acts as a scattering center and, hence, the total intensity in a given direction of the diffracted beam depends on how completely the contributions from individual atoms are in phase. By trial and error, the arrangement of atoms that best accounts for the observed intensities of reflections is selected.

Two experimental methods are widely used for studying X-ray diffraction. The first method uses a single crystal of the compound, which is mounted and rotated about a crystal axis. A monochromatic X-ray beam impinges on the crystal as it rotates, bringing successive sets of crystal planes into reflecting positions. The reflected beams are recorded as a function of crystal rotation, and the positions and intensities are recorded. From these data, a complete structure can be derived.

The second method is used when only a crystalline powder is available. The powder is placed in a beam of monochromatic X-rays. Each particle of the powder behaves as a tiny crystal or assemblage of smaller crystals. They are oriented randomly with respect to the incident beam. Some of the crystals will be correctly oriented so that (100) planes can reflect the beam. Others will be correctly oriented for (110) reflections, etc. Each and every set of lattice planes will be capable of reflection and, hence, the mass of powder is equivalent to a single crystal rotated not about one axis, but about all possible axes. Older instruments used film techniques to record the data obtained from single crystals or polycrystalline samples. However, today almost all X-ray diffractometers utilize scintillation counters and record the diffraction data on paper or magnetic media.

X-Ray powder diffraction is used as a standard technique for the identification of phases and the determination of their composition. The latter can be accomplished when there is a systematic variation in cell sizes as a function of composition. Precise cell dimensions can be calculated from data obtained by the step counting of the goniometer over the peaks of the sample and those of an internal standard. The identification of an unknown requires obtaining a diffraction pattern recorded with a Debye–Scherrer camera, Guinier camera, or diffractometer. From the diffraction pattern, the plane spacing d corresponding to each line of the pattern can be determined.

Relative intensities of the peaks on a film are usually estimated by eye. The intensity of a peak on a diffractometer recording is often taken as the maximum height above the background, but a more reliable technique is to integrate the area under the peak. Division of this area by the peak height yields the "integral width" of the peak, and this width can be used in the Scherrer equation to ascertain the average crystallite size in a powder sample. The equation can be written as

$$D = \frac{k\lambda}{\beta \cos \theta}$$

where

D = crystallite size
k = shape factor
β = integral breadth or half-height breath of peak at θ radians
λ = X-ray radiation wavelength
θ = the Bragg angle (half of the peak position angle).

In many studies, a shape factor value of 0.9 and the value of 1.5404 Å for the wavelength of $CuK\alpha_1$ radiation were used. Variability in sample crystallite sizes may cause a number of irregularities and asymmetric broadening of the diffraction line. Hence, the half-height breadth is not an adequate measurement of the broadening. The integral breadth, rather than the half-height breadth, is needed for each diffraction peak. The integral breadth is the total area under the curve (which is proportional to the total scattering) divided by the height of the peak. The details of extracting the true breadth from the observed breadth are discussed in Refs. 2–4.

When the major phase consists of a solid solution between ions of sufficiently different atomic number, then the relative peak intensities vary with composition and can be used to characterize it. In addition, variable structure parameters (e.g., the anion position in the spinel unit cell) and the distribution of ions (including vacancies) among distinct crystallographic sites can be determined (with varying degrees of reliability) from a detailed fitting of parametrized theoretical models to the observed relative intensities.

Modern diffractometers are equipped with programs that not only list all of the *d* spacings and their relative intensities, but also match these values with those of all known and recorded compounds. The best matches are then presented for final selection of the most appropriate candidate. If this procedure is done manually, then the closest match of the strongest three reflections is made with cards from the Powder Diffraction File published by the Joint Committee on Powder Diffraction Standards (JCPDS). The most appropriate card or cards are chosen and the *d* values for all of the reflections, as well as their intensities, are compared to the values obtained for the unknown. When full agreement is obtained, then the identification can be made.

Recently, studies have indicated the importance of Guinier and Gandolfi cameras in addition to those using the standard packed holder, smear-type, or spindle powder mounts. The Gandolfi permits the propagation of an entire diffraction pattern from a small single crystal. The Gandolfi camera is designed for generating powder diffraction patterns from single crystals as small as 30 μm. This can be accomplished without regard to orientation and without having to remount the crystal. This is achieved by applying rotation about a second axis which is inclined at 45° to an ordinary Debye–Scherrer camera. This results in a randomization of the orientation and generates a complete powder pattern. The camera can also be used as a Debye–Scherrer camera, and is designed for vacuum or atmosphere operation. It is particularly useful when a specimen cannot be ground up or destroyed.

The Guinier deWolff camera allows the detection of small amounts of phases that can be missed by other powder methods. The Guinier camera is a cylindrical camera with the specimen and film arranged on the surface of the cylinder. The camera was developed in the late 1930s and is a combination of a focusing monochromator and a focusing camera. The Guinier deWolff camera is a set of two (or four) Guinier cameras stacked one above the other and separated by baffles. Hence, patterns from two (or four) different specimens can be registered simultaneously on the same piece of film. This arrangement is possible because the beam from the monochromator can be split into two (or four) beams. Compared to a Debye–Scherrer camera of the same size, a Guinier camera provides a much clearer pattern with twice the resolution and about the same exposure time. However, any one Guinier camera covers only a limited range of 2θ. It is best suited to the examination of particular parts of complex patterns.

In general, X-ray identification is capable of detecting approximately 2% of a foreign phase under favorable circumstances. X-Ray diffraction and observation of optical characteristics by standard petrographic techniques are used to identify the phase present, while petrographic techniques alone are often capable of greater subtlety in detecting traces of foreign phases. However, ordinary microscopic observations, including examination between crossed polarizers, are often neglected by those who tend to use X-ray diffraction as the mainstay of phase identification. It should be emphasized that optical examination combined with etching and similar metallographic techniques are often capable of shedding light on phase homogeneity in both metallic and nonmetallic crystalline solids.

e. Neutron Diffraction

If a small opening is made in the wall of a nuclear reactor, a beam of neutrons can be obtained. A monochromatic beam, with a single energy, can be obtained from a single crystal, and this beam can be used in neutron diffraction experiments. Diffraction experiments can be performed with a spectrometer in which the intensity of the beam diffracted by the specimen is measured with a proportional counter filled with BF_3 gas.

The intensity of neutron scattering varies irregularly with the atomic number of the scattering atom. Some light elements, such as carbon, scatter neutrons more intensely than some heavy elements, such as tungsten. Whereas X-rays cannot be used to determine the position of hydrogen atoms in a compound, neutron diffraction can readily accomplish this. Neutron diffraction can also distinguish in many cases between elements which differ by one atomic number, whereas X-ray diffraction cannot distinguish the positions of elements which scatter X-rays with almost equal intensity. In addition, neutrons have a small magnetic moment. If the scattering atom also has a net magnetic moment, interaction between the two will modify the total scattering. If the specimen has an ordered arrangement of atomic moments (antiferromagnetic, ferrimagnetic, or ferromagnetic), neutron diffraction can provide knowledge of both the magnitude and directions of these moments.

f. Thermal Analysis

Thermal analysis techniques are used to measure the physical and reactive properties of materials as functions of temperature. Conventional thermal analysis depends on the observation that a phase change produces either an absorption or evolution of heat. For the case where a pure melt solidifies, the solidification will start at a characteristic temperature. The freezing process is accompanied by an evolution of heat and the rate of this process is related to the rate of heat loss. In Fig. 2.1a can be seen the cooling curve of a pure substance. The horizontal line represents the constant temperature during which solidification occurs. If two substances are completely immiscible in the solid state and do not form a compound, the cooling curve of their solution can be represented by Fig. 2.1b. In this case, the melt will cool until one component begins to solidify (point *A*). The temperature will continue to fall (but less rapidly) because freezing of a solution does not occur at constant temperature. The decrease in temperature will continue until the eutectic temperature *B* is reached and this temperature will remain constant until solidification is complete (point *C*).

Differential thermal analysis can be used in place of the simple cooling-curve method. A differential method is used to detect heat absorbed or evolved during a phase change. Two identical chambers are symmetrically located in a furnace and the precise temperature of each chamber is continuously ascertained. The sample is placed in one chamber and an inert material, e.g., α-Al_2O_3, in the other. The furnace is heated at a controlled uniform rate and the temperature difference is recorded as a function of time. If reactions occur in the sample chamber, peaks will be obtained in the plot of temperature differential versus furnace temperature. The peaks from exothermic and endothermic reactions will appear in opposite directions. This is clearly shown in Fig. 2.2.

In thermogravimetric analysis, the weight of a heated sample is measured as a function of temperature and time. The apparatus consists of a sample container enclosed in a vertical tube furnace and suspended from a sensitive balance.

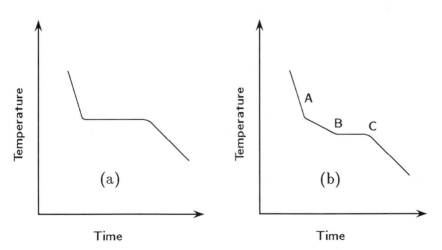

Figure 2.1. Cooling curves for (a) a pure substance and (b) a mixed system of two immiscible substances.

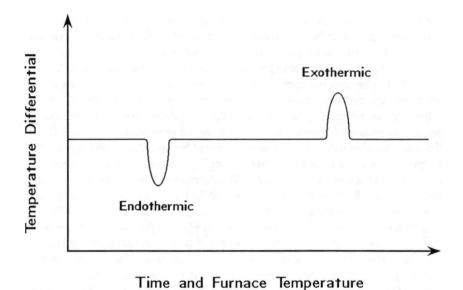

Time and Furnace Temperature

Figure 2.2. Warming plot from differential thermal analysis showing endothermic and exothermic reactions.

Changes in weight are recorded as a function of temperature and time. Volatile products result in a measurable weight loss, and oxidation or adsorption of gases in a weight gain. Figure 2.3 indicates the results obtained on heating manganese (II) oxalate. The dihydrate is stable to 54°C. Complete dehydration occurs at 100°C and decomposition to CO and CO_2 above 214°C yields MnO. At higher temperatures, the MnO gains weight as it is oxidized to Mn_3O_4.

g. Wet Chemistry

The analytical chemist has used wet chemistry for years and it is still a widely used technique for establishing composition. Where applicable, classical analytical techniques are still the best choices for determining the concentration and stoichiometry of the major phase. In this connection, wet chemical analysis, including gravimetric, volumetric, and electrochemical techniques, can be used. However, some of the instrumental methods, e.g., X-ray fluorescence, have the advantage of rapidity of execution. Recently, the automation of wet chemical techniques for analysis of solids has been developed and are similar to those used in routine clinical laboratories.

The exact determination of stoichiometry when precise results are required almost always depends on indirect measurement and on a "logical chain of reasoning," which is often not tenable. Conductivity and mobility measurements together with assumptions as to the origin of donors or acceptors sometimes allow

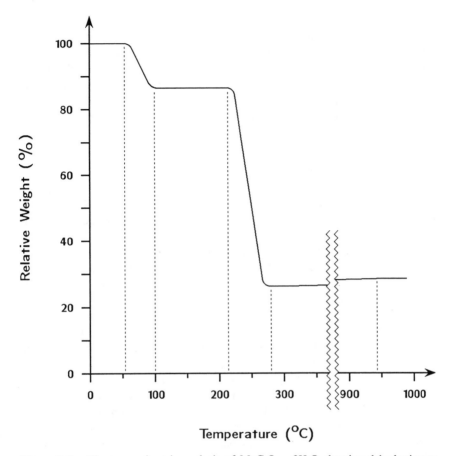

Figure 2.3. Thermogravimetric analysis of $MnC_2O_4 \cdot 2H_2O$ showing dehydration to MnC_2O_4 at 54–100°C, decomposition to MnO at 214–280°C, and oxidation to Mn_3O_4 at elevated temperatures.

estimates of stoichiometry to be made in semiconductors. Similarly, precise lattice parameters and pycnometric density determinations provide some means of estimating vacancy concentration and, in some cases, stoichiometry.

Once the stoichiometry has been determined, it is conventional to assign "formal valences" of multivalent elements in a compound in order to preserve electrical neutrality. This procedure presents difficulties when more than one multivalent element is present in the same compound or where several valency assignments are possible for a given element in the compound. If it is possible to get the compound into solution without altering the valence, the classical wet chemical and electrochemical methods are quite precise (0.01%) and permit the determination of "valence state." For iron and some other elements, a specialized technique of Mössbauer spectroscopy is applicable with a precision of about 0.1%. Improvements in the precision of Mössbauer measurements are likely. High-resolution X-ray spectra also give information on valence which is applicable to most elements. In actual practice, the valence is often deduced from the color of the compound. Absorption spectroscopy is useful in particular circumstances, but it will probably be limited to the determination of the valence of trace elements. Similarly, electron spin resonance is applicable to elements present in trace concentrations.

h. Spectroscopic Methods

Chemists are most likely to make direct measurements of electronic structure in the ultraviolet and visible region, of vibrational spectra in the infrared region and by the Raman effect, and of nuclear magnetic resonance in the radio frequency region. Ultraviolet or visible radiation is commonly required to produce excited electronic states of molecules. The lower energy infrared is sufficient for vibrational transitions, and pure rotational transitions require still lower energy microwave and radio frequency regions.

Spectroscopic techniques give information concerning the coordination number and site symmetry of a species. Unlike X-ray methods, these techniques can be used to study the structure of amorphous materials such as glasses and gels, as well as crystalline solids.

The equipment used to study ultraviolet, visible, and infrared spectra is similar. A sample is placed in a beam of radiation and the transmitted energy is detected as a function of frequency. For Raman spectroscopy, the sample is illuminated with strong monochromatic radiation, and some of the reemitted quanta are found to have gained or lost energy corresponding to the energy of the fundamental vibrational modes. The spectrum obtained consists of a very strong line corresponding to the incident radiation and a number of additional lines whose energy differences from the primary line give the energies of vibrational transitions of the molecule.

Transitions of outer shell electrons fall in the general wave number range of 1×10^4 to 1×10^5 cm^{-1}. The position of an absorption band corresponds to the energy of the transition; its intensity depends on the nature and quantity of the absorbing material. The relations between the absorbance A, the pathlength ℓ, and the concentration c is given by the Beer–Lambert law

$$A = kc\ell$$

where k is a constant for the material. For $\ell = 1$ cm and the concentration expressed in moles per unit volume, k becomes e, the molar extinction coefficient. Information about the energy levels of the valency orbitals is obtained from such electronic spectra. This has wide application in the study of the transition metals. Beer's law allows for the use of absorbance measurements in quantitative analysis.

Vibrational modes of a molecule are excited by the absorption of quanta whose energy lies in the infrared region of the spectrum (4000 cm^{-1} or less). Vibrational spectra are also detected in Raman scattering, and the two techniques are complementary. Other resonance methods have been applied to the study of solids such as Mössbauer, nuclear magnetic resonance, and electron spin resonance. These have, in the past, been of limited value because they are applicable to the study only of specific elements. Hence, the number of systems which can be studied are also limited. Mössbauer spectroscopy is referred to in regard to several of the iron-containing systems discussed in the latter sections of this book. Hence a brief description of this method will be given.

Mössbauer spectroscopy is concerned with transitions that occur inside atomic nuclei. The technique measures the resonant absorption of a γ-ray by a nucleus held rigidly within a crystal structure. If an isotope, such as ^{119}Sn, has a metastable excited state that transforms to the normal ground state of the nucleus by emitting a γ-ray, the excited state can be used as a source and the ground state as target for the emission and reabsorption of the γ-ray. This effect is the basis for Mössbauer spectroscopy. The monochromatic γ-rays absorbed by the sample can have their energy varied by use of the Doppler effect. The absorption of the γ-rays can be recorded as a function of energy. This gives a spectrum that consists of a number of peaks. The conditions for absorption of the γ-rays depend on the electron density at the nucleus, and the number of peaks obtained is related to the symmetry of the compound. The elements most commonly studied are Fe, Sn, and I, and the technique can be used to identify the formal oxidation state and site preference of iron in various compounds.

B. Characterization of Minor Phases and Impurities

Most inorganic impurities can be detected down to about 10 ng/g, though for quantitative determination, this impurity level cannot be attained. The sensitivity of quantitative analysis by most instrumental methods is frequently limited by a lack of suitable standards.

Emission spectroscopy is probably still the most generally suitable technique of the survey methods. It is used widely for the characterization of solids, liquids, and gases, and has the capability of detecting by dc arc excitations up to 70 elements ranging from 0.1 to 100 ppm. Determination of nonmetallic elements is also possible with emission spectroscopy, but requires special techniques which are rarely used. The nature of the spectrum of light emitted by each element when it is heated or burned has been the basis for identification of the elements for well over 100 years. This method still is among the most sensitive means for detection of small traces of impurities. Techniques have been developed that make it a fairly accurate method of quantitative chemical analysis, supplementing, and many times supplanting, classical wet chemical methods. However, the optical spectra of most elements are very complicated. They can contain even thousands of lines. In principle, this is not a limitation, but the usefulness of spectrochemical analysis is often limited in practice by difficulties in the identification of a mixture of components where the spectrum is crowded.

X-Ray spectroscopy can also be employed for survey analysis of impurities and has the advantage of being nondestructive. Lower limits for the detection of transition metals are rarely better than 10 to 100 ppm, and the elements of the first period cannot be detected at low concentrations. This method is based on X-ray instead of the optical spectra emitted by the elements. It has characteristics similar to those of the optical method and it has additional features (e.g., that of being nondestructive). The method identifies the elements and their concentrations by means of the wavelengths and intensities of lines in their characteristic X-ray spectra. The technique can be applied to all kinds of specimens, solid, powder, crystalline, or amorphous. Cases of overlapping lines of different elements are not frequent and methods are available for separating or distinguishing them. The X-ray spectra are usually excited by way of fluorescence, induced by the continuous or characteristic radiation of an X-ray tube with tungsten or other target. The X-ray spectra are analyzed by diffraction from a single crystal. A Geiger, proportional, or scintillation counter is used for measuring the reflected rays. Hence, the instrumentation is very similar to that used in X-ray diffractometry.

Electrical measurements (see Chapter 3) are useful for determining the total content of electrically active impurities in conductors and semiconductors. Also, the shape of the freezing curve can give considerable information on the total impurities in a material with a suitable melting point. However, this technique is of little importance for most metals, which melt at high temperature, or for materials which decompose on melting.

a. Limitation of Analytical Methods

Most of the analytical techniques used to determine impurities have serious limitations because of poor detection limits, the need for standards, and the hazard of contamination, particularly when preliminary handling of the sample is required. Activation analysis is suitable for trace characterization since it permits most sampling to be done after the activation step, and thus contamination problems are minimized. However, this method requires costly equipment and considerable effort. Spectroscopic methods, such as emission, mass, or X-ray, provide for the determination of many elements in small absolute amounts, but are limited in the amount of sample that can be analyzed. They are also useful methods when used after separation or preconcentration, as a group, of several impurities. These techniques are also limited by the availability of suitable standards.

The various techniques that can give information on valence and location of ions all suffer from relatively poor sensitivity (perhaps 1000 ppm) and are thus more suitable for the characterization of a major phase than of the impurities. Magnetic resonance techniques such as nuclear magnetic resonance (NMR) and electron spin resonance (ESR) have low limits of detection but both are limited in applicability and are difficult to quantify.

References

1. R. A. Adams, Ph.D. thesis, Brown University, 1973.

2. Characterization of Materials Prepared by the Committee on Characterization of Materials. Materials Advisory Board, National Academy of Sciences/National Research Council, Washington, D.C. March 1967.

3. A. K. Cheetham and P. Day (eds), *Solid State Chemistry Techniques*, Oxford University Press, Oxford, 1987.

4. B. D. Cullity, *Elements of X-Ray Diffraction*, 2nd ed. Addison-Wesley, Reading, MA, 1978.

Additional References

E. M. Chamot and C. W. Mason, *Handbook of Chemical Microscopy*, 3rd ed. John Wiley, New York, 1958.

P.A. Cox, *The Electronic Structure and Chemistry of Solids*, Chap. 2, Oxford University Press, New York, 1987.

W. J. Croft, *Ann. N.Y. Acad. Sci.*, **62**, 464 (1956).

H. P. Klug and L. E. Alexander, *X-Ray Diffraction Procedures*, John Wiley, New York, 1974.

C.N.R. Rao and J. Gopalakrishnan, *New Directions in Solid State Chemistry*, Chap. 2, Cambridge University Press, New York, 1986.

J. P. Sibilia (ed.), *A Guide to Materials Characterization and Chemical Analysis*, VCH, New York, 1988.

F. D. Snell-Hilton, *Encyclopedia of Industrial Chemical Analysis*, **3**, 796 John Wiley, New York, 1966.

A. F. West, *Solid State Chemistry and Its Applications*, John Wiley, New York, 1987.

Problems

P2.1. Each crystalline solid gives a characteristic X-ray powder diffraction pattern that can be used as a fingerprint for its identification. Discuss the reasons for the validity of this statement and indicate how two solids with similar structures, e.g., NaCl and NaF, can be distinguished by their powder patterns.

P2.2. Show by means of qualitative sketches the essential differences between the X-ray powder diffraction patterns of (a) 1:1 mechanical mixture of NaCl and AgCl and (b) a sample of (a) that has been heated to produce a homogeneous solid solution.

P2.3. Discuss the techniques that you would use to determine the following:
   a. Whether or not MnO possessed a magnetically ordered superstructure
   b. Whether $YBa_2Cu_3O_{6.8}$ was a superconductor
   c. The nature of a surface layer of Al_2O_3 on a piece of aluminum metal
   d. The coordination of Cr^{3+} in a crystal of ruby.

P2.4. A sample of aluminum oxide Al_2O_3 was shown by chemical analysis to contain 10 atomic percent Cr^{3+}:
   a. What effect would the Cr^{3+} have on the X-ray pattern if the Cr^{3+} substituted for Al^{3+} in the crystal structure of Al_2O_3?
   b. How would you determine the limit of solubility of Cr^{3+} in Al_2O_3?
   c. How would the spin only moment of Cr^{3+} differ between a sample containing 10% substitution of Cr^{3+} for Al^{3+} and one containing 5% substituted Cr^{3+}? Explain the reason for this difference if one is observed.

P2.5. Choose any compound we have studied:
   a. What are the methods for preparing polycrystalline samples, single crystals, thin films.
   b. What is the structure of the compound you have chosen.
   c. How does the structure and stoichiometry affect either an electronic, magnetic, or optical property?

P2.6. Select a catalytic process that utilizes a solid state material. Discuss:
   a. Optimization of the surface area of the catalyst and affect on the process.
   b. Choice of precursor based not only on ability to give highest activity but cost of precursor and processing must be considered.
   c. Avoidance of hazardous procedures.
   d. Reproducibility of activity when process is transferred from laboratory to production scale.

It is suggested that for specific catalysts and processes, the student is referred to books such as B. C. Gates and G. C. A. Schmit, *Chemistry of Catalytic Processes*, McGraw-Hill, New York, 1979.

3

Electrical Properties of Semiconductors

Semiconductors are compounds that show electrical conductivity intermediate between that of metals and insulators. One of the best criteria for differentiating between metallic and semiconducting behavior is to observe the variation of conductivity with temperature. The measured electrical resistance of metals is due to the scattering of carrier electrons by interaction with a nonperiodic lattice potential. This can occur either by the presence of lattice defects such as vacant sites, impurity atoms, or grain boundaries or by thermal motion of the lattice. For metals, the carrier concentration does not change with increasing temperature, but lattice motion does increase. Hence, the conductivity, which depends on both the number of carriers and their mobility, will decrease with increasing temperature. For semiconductors, the number of carriers increases rapidly with temperature. Hence, despite the increased lattice motion, conductivity will increase with temperature.

Semiconductors may be divided into those that show intrinsic conduction and those which are extrinsic. Intrinsic behavior is shown by so-called "perfect" crystals, whereas extrinsic (or impurity type) semiconductors contain foreign atoms that are present either substitutionally or interstitially in the host crystal.

In the study of transport properties of solids, two types of charge carriers can be present within the crystal, namely, electrons and holes. Actually, as can be seen in Figs. 3.1a and 3.1b, all conduction is related to the movement of electrons. However, two processes are possible. In Fig. 3.1a, the excess electron moves from site to site and this "electron flow" gives rise to n-type behavior; in Fig. 3.1b where there is a deficiency of electrons, each electron jumps in turn so that it appears that the hole is moving continuously. This apparent movement of holes can be described as hole conduction, and is called p-type behavior. The concept of positive hole conduction is important in the study of semiconductors because of the fundamental difference in its resulting behavior compared to electron conduction.

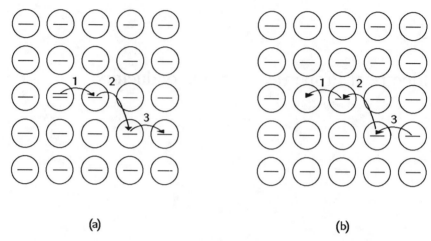

(a)   (b)

Figure 3.1.   Conduction due to (a) an excess and (b) a deficiency of electrons.

The electronic behavior of semiconductors is usually explained by band theory. A simple "free atom" model may be used to interpret the properties of materials such as silicon. Two atoms are brought together so that their wave functions overlap to form two allowed energy states $\psi_a + \psi_b$ and $\psi_a - \psi_b$ and each state can contain two electrons. The energy of these combined states as a function of internuclear distance is given in Fig. 3.2. If an infinite number of atoms are used, the number of allowed energy levels becomes large and form bands (Fig. 3.3).

A general overview of the properties of semiconductors was presented above. A somewhat more detailed treatment will be required to understand the subtle differences between compounds of similar composition or structure. The original model of energy bands arose from considering the effect of a periodic structure on the wave functions of otherwise free electrons. The existence of forbidden energy gaps between permitted bands resulted from the continuity required by periodic boundary conditions. But the complicated details of the band structures of the materials discussed in this book require that one start with atomic wave functions. These are modified by any crystal electric field present at that site and form anion–cation, cation–cation and anion–anion molecular orbitals. Bands are formed by linear combinations of the resulting hybridized wave functions.

These band wave functions, of course, consist of linear combinations of molecular orbitals from every pair of interacting ions, each combination being associated with a different energy value within the energy band. However, the properties of the material can be adequately understood, and more readily discussed, in terms of the parentage of these combinations, i.e., the interaction between individual pairs of ions, with the knowledge that the resulting energy level will be spread out into a band of levels that will be broad if the overlap of atomic orbitals is strong and narrow if it is weak. The strength of the overlaps between the hybrid-

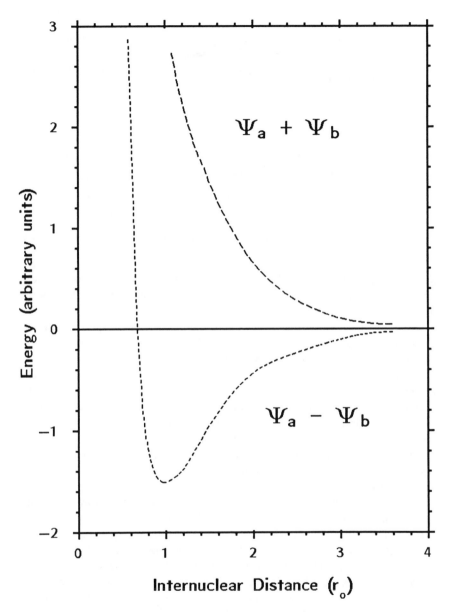

Figure 3.2. Energy vs internuclear distance for overlap of ψ_a and ψ_b.

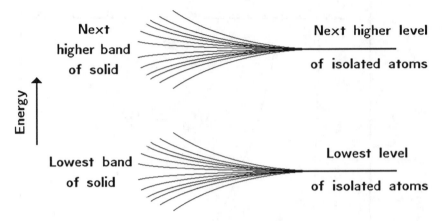

Figure 3.3. Splitting of the discrete energy levels of isolated atoms into bands.

ized orbitals of two neighboring ions will depend on the orientation of those orbitals with respect to the crystal structure, upon the bond angle and upon the bond distance.

The case of transition metal d-electron functions deserves special mention. These wave functions are, in fact, multi-electron functions. However, their relative energies and properties can be described in terms of a one-electron model in which the wave functions and energy levels of an ion with a single d-electron in the crystal electric field are populated successively from the bottom by the electrons belonging to the transition metal. In other words, the d-electron functions can be described in terms of their parentage. The bands formed from molecular orbitals involving these d-electron functions overlapping with neighboring ions are similarly populated from the bottom by the available electrons. Specific examples of this treatment involving d-electrons will be presented in the body of this text.

In the simpler case of silicon (no d-electrons), as electron wave functions overlap, the $3s^2 3p^2$ states broaden to give a bonding and an antibonding band. The s and p states have mixed together in silicon to give hybrid states, namely the sp^3 tetrahedral orbitals of silicon. Since there are four covalent bonds, the sp^3 states contribute four electronic states per atom, and in a crystal of silicon, all of the bonding states are filled with the available electrons. The energy-band structure of silicon can be represented as shown in Fig. 3.4. The highest filled energy level lies at the top of the valence band, and the lowest empty level lies at the bottom of the conduction band. According to statistical mechanics, no conduction is possible at 0 K, since all of the available states in the lower band are filled and all of the upper band states are empty. However, as the temperature is raised, electrons are thermally excited from the valence band into the conduction band, leaving holes which can conduct in the valence band, and electrons to conduct

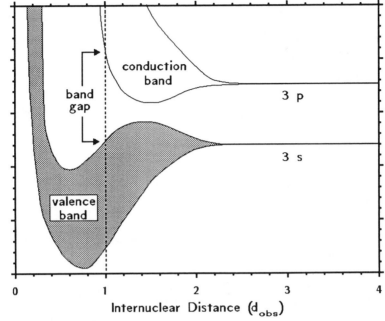

Figure 3.4. Energy band structure of silicon.

in the conduction band (Fig. 3.5). This process gives rise to equal numbers of positive and negative charge carriers which is characteristic of intrinsic conductivity. Pure silicon is an intrinsic semiconductor and the conductivity of pure silicon is plotted in Fig. 3.6 as log σ vs $T^{-1}(K^{-1})$. From this plot, $\sigma = \sigma_o e^{-\varepsilon/(2kT)}$ and log σ = log σ_o - $\varepsilon/(2kT)$. From the slope of the plot, the value for the thermal band gap ε may be determined.

If a crystal is not perfect, then extrinsic conduction arises from atomic defects. There are four major types of atomic defects that give rise to extrinsic conductivity:

1. Lattice sites that should be occupied in a perfect crystal but are vacant.

2. Interstitial atoms which occupy sites that are not permitted in a perfect crystal.

3. Misplaced atoms or atoms that are on incorrect lattice sites, e.g., an anion on a cation site or vice versa.

4. Atoms with higher or lower than usual valence residing on lattice sites.

Any of these defects may ionize and give rise to conductivity. If, for example, an oxygen is missing from the crystal lattice of a metal oxide (MO), then there must be two electrons trapped in the neighborhood of the vacancy to preserve

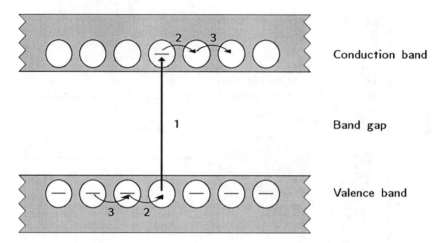

Figure 3.5. Promotion of electrons from the valence band to the conduction band in an intrinsic semiconductor.

electrical neutrality (Fig. 3.7). These electrons may be dissociated from the defect and become conducting electrons. Hence, the oxygen vacancies are called donors because they can donate electrons to the conduction band. This process is shown in Fig. 3.8.

The presence of metal vacancies in a crystal traps positive charges in the vicinity of the defects so that electrical neutrality is preserved. The positive charge may be separated from the defect, and although each electron jumps in turn, it is the positive charge or hole that appears to be moving. This type of hole migration gives rise to p-type conduction. In Fig. 3.9, it can be seen that in this case the vacancies accept electrons and hence the defects are known as acceptors. P-type conduction takes place in the valence band as is shown in Fig. 3.9.

Electrons and holes differ not only in having opposite charges, but also with regards to differences in mobility. Hence the total conductivity, σ, of a substance is given by the expression

$$\sigma = ne\mu_- + pe\mu_+$$

where n and p are the numbers of positive and negative charge carriers per unit volume, e is the electronic charge, and μ_+ and μ_- are the mobilities of the positive and negative charge carriers expressed as cm^2/V-sec in the c.g.s. system. The units of σ are usually $\Omega^{-1} cm^{-1}$.

The conductivity of a cylindrical sample can be determined by measuring its resistance and then multiplying the result by a geometrical factor. This is given as

$$\sigma = 1/\rho = \ell/RA$$

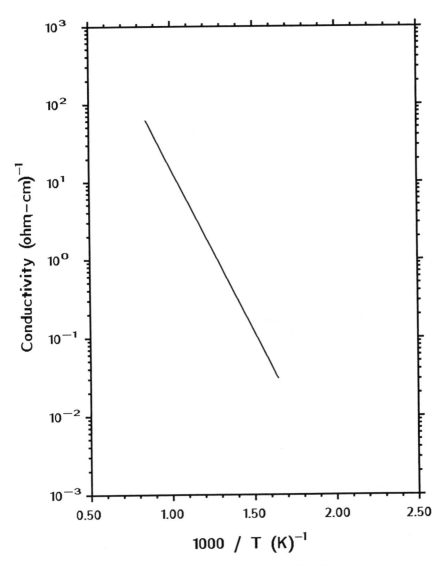

Figure 3.6. Electrical conductivity of pure Si plotted against inverse temperature.

where ℓ is the length of the sample, A is the cross-sectional area, and R is the resistance. This measurement can be done by either a two-probe or a four-probe method. In the two-probe technique, the resistance is measured across the sample with an ohmmeter. The main disadvantage to this method is that contact resistance is included as part of the sample resistance, introducing serious errors in the results.

$$M^{++} \qquad O^= \qquad M^{++} \qquad O^=$$

$$O^= \qquad M^{++} \qquad O^= \qquad M^{++}$$

$$M^+ \qquad\qquad\qquad M^+ \qquad O^=$$

$$O^= \qquad M^{++} \qquad O^= \qquad M^{++}$$

Figure 3.7. An oxygen vacancy in an MO lattice.

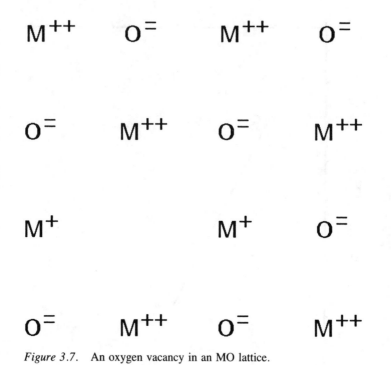

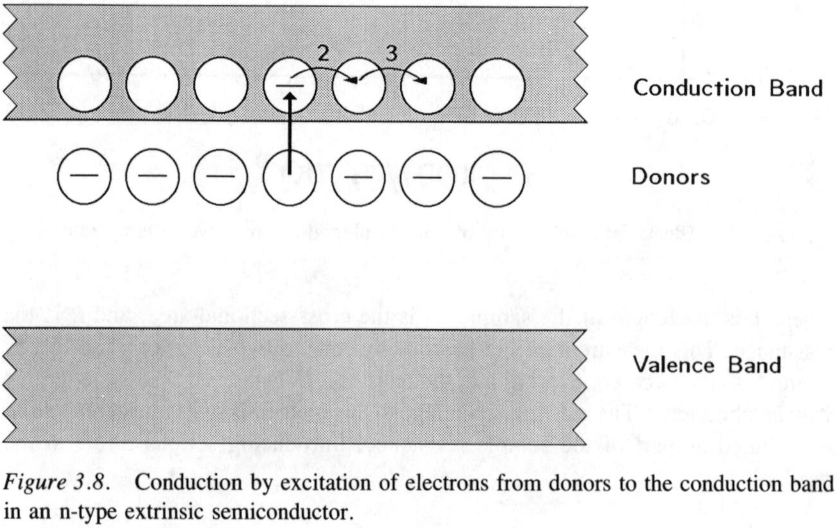

Figure 3.8. Conduction by excitation of electrons from donors to the conduction band in an n-type extrinsic semiconductor.

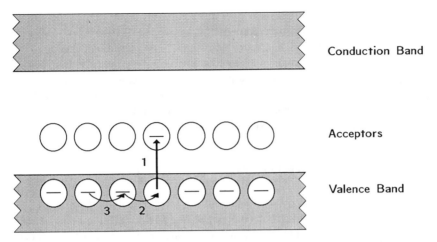

Figure 3.9. Conduction by excitation of electrons from the valence band to acceptors in a p-type extrinsic semiconductor.

The four-probe method is better since two probes are used to pass current through the sample and two other probes are connected to a voltmeter which has a high input impedance. Measurement of resistivity by means of the four-probe method is illustrated in Fig. 3.10. R_1 is a standard resistor that is used to determine the current flowing through the sample $I = V_1/R$. The resistance of the sample may then be calculated from $R_s = V_2/I$. The length is now the distance between the probes connected to V_2, which are called the voltage probes. The advantage to this method is that the high impedance voltmeter sees only the sample and not the current contacts. Therefore, errors due to contact resistance are eliminated.

More recently, another method, known as the van der Pauw technique, has

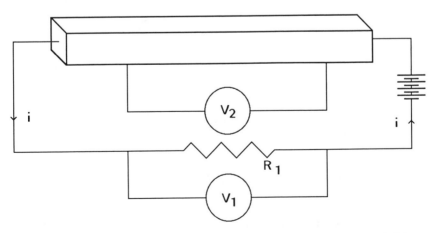

Figure 3.10. Schematic diagram of the four-probe method for measuring resistivity.

been developed to measure resistivity of any sample shape, as long as the sample is thin and does not have sharp corners.

In this technique (Fig. 3.11), four small contacts are placed on the edge of the sample, preferably approximately equally spaced. A current is then passed through two adjacent leads and the voltage is measured across the other two leads as described for the four-probe method. Then, one of the current and voltage leads is interchanged so that the current leads are still adjacent, and another measurement is taken. The resistivity is determined for the general case by the following equation:

$$\rho = [\pi t f/(2\ell n2)](R_{MN,OP}+R_{NO,PM})$$

where t is the thickness and f is a factor that is a function only of the ratio $R_{MN,OP}/R_{NO,PM}$, and can be obtained from a plot of f vs $R_{MN,OP}/R_{NO,PM}$. Such a plot is given in van der Pauw's paper (1). $R_{MN,OP}$ and $R_{NO,PM}$ are determined from the voltage and current measurements made from contacts at M, N, O, and P (see Fig. 3.11). The advantage to this method is that the length or area of the sample does not have to be known, and almost any shape of sample can be used.

To determine the number of carriers, it is also necessary to know the Hall voltage. The Hall voltage is defined as the voltage generated in the z direction when a current is passed through a sample in the x direction in the presence of a magnetic field along the y direction as shown in Fig. 3.12 for a p-type semicon-

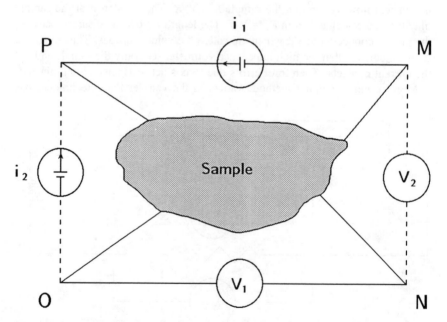

Figure 3.11. Schematic diagram of the van der Pauw method for measuring resistivity.

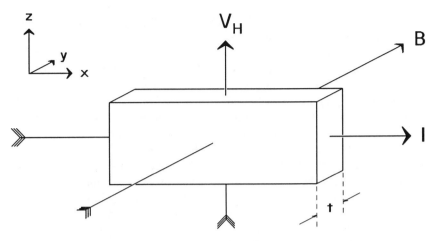

Figure 3.12. Geometry for measuring the Hall effect.

ductor. The Hall voltage, V_H, is proportional to the current and the magnetic field and is inversely proportional to the thickness of the sample in the direction of the magnetic field. The proportionality constant, R_H (10^8 cm^3/C), is called the Hall coefficient, and is given by

$$R_H = 10^8 V_H t/(IB)$$

where t is the thickness in the y direction in cm and B is the magnetic field strength in gauss. The Hall coefficient for a semiconductor with both holes and electrons as carriers is given by the expression

$$R_H = [\sigma_p^2/(pe) - \sigma_n^2/(ne)]/\sigma^2$$

where σ_n and σ_p are the conductivities of the negative and positive charge carriers, respectively. The sign of the Hall coefficient is determined by the relative contributions of holes and electrons to the conductivity.

It is evident that the measurement of the Hall effect may be used to determine directly the number of charge carriers for a semiconductor which has only one type of charge carrier. Substitutions into the equation for the conductivity of a single type of charge carrier

$$\sigma_n = ne\mu_n \quad \text{or} \quad \sigma_p = pe\mu_p$$

give

$$R_H = 1/pe = -1/ne$$

so that the mobility is also obtainable directly from Hall and conductivity measurements. When both electrons and holes contribute to the Hall effect, the consequences can be complex.

For numerous applications, dopants are added to the Group IV semiconductor to obtain the desired number and type of carriers. The semiconductors are then extrinsic. The doping can be achieved either by the addition of a Group III or Group V element of the Periodic Table. In pure silicon, each silicon atom has four valence electrons and is coordinated to four other silicon atoms. However, gallium has only three valence electrons. Hence, each bond involving Ga–Si must be deficient by one electron. The single electron gallium–silicon bond forms a discrete level above the valence bond and is known as an acceptor level because it can accept an electron. Hence, electrons from the valence band can be thermally excited into acceptor levels. The positive holes left behind in the valence band give rise to p-type semiconduction.

The substitution of an arsenic atom for silicon results in an excess of one electron for each arsenic–silicon bond. The extra electron occupies a discrete level below the bottom of the conduction band. Electrons from donor levels can be thermally excited into the conduction band.

The doping of silicon may be related to the ionization of a weakly dissociated solvent such as water. The dissociation of water is

$$H_2O \rightleftharpoons H^+ + OH^-$$

and

$$K_w = [H^+][OH^-]$$

where the square brackets denote concentration:

$$[H^+] = [OH^-] = K_w^{1/2}$$

Similarly for $Si \rightleftharpoons e^- + h^+$, where an electron is excited to a conduction band leaving a hole in the valence band,

$$K_i = [e][h]$$

and for pure Si,

$$[e] = [h] = K_i^{1/2}$$

In the above expression, $[e] = n$ and $[h] = p$. The addition of a Group V element to silicon as impurity atoms results in the formation of donor levels capable of contributing electrons into the conduction band. When electrons are excited from the donor level into the conduction band, they are dissociated from their donor sites. The equilibrium expression

$$K_i = [e][h] = np$$

is valid but $n >> p$. The introduction of donor atoms into silicon can be related to the addition of a weak base to water. In the latter case, there is an increase in OH⁻ concentration and a decrease in the H⁺ concentration.

If instead of a Group V element, a Group III element, e.g., B, is introduced into silicon, a new level called an acceptor level would be introduced into the band gap somewhat above the top of the valence band and electrons can be promoted from the valence band to the acceptor level. The positive holes thus formed in the valence band give rise to p-type conductivity. This situation is analogous to the dissociation of a weak acid in water, which increases H⁺ ions and decreases OH⁻ ions.

For an intrinsic semiconductor, the equilibrium concentration of electrons in the conduction band, or holes in the valence band, can be given by the following expressions readily derived from Fermi–Dirac statistics

$$n = N_c \exp[(E_F - E_c)/kT] \quad \text{and} \quad p = N_v \exp[(E_v - E_F)/kT]$$

where

N_c = density of states in the lowest part of the conduction band
E_c = energy at the same level
E_v = energy level at the top of the valence band
N_v = density of states in that region (the total number of quantum states per unit energy)
E_F = Fermi energy or level (for this energy, there is an equal probability of finding an electron or hole).

For an intrinsic semiconductor, $n = p$. Hence

$$N_c \exp[(E_F - E_c)/kT] = N_v \exp[(E_v - E_F)/kT]$$

Taking logarithms and rearranging,

$$E_F = (E_v + E_c)/2 + (1/2)kT \ln(N_v/N_c)$$

When $N_v = N_c$, it follows that

$$E_F = (E_v + E_c)/2$$

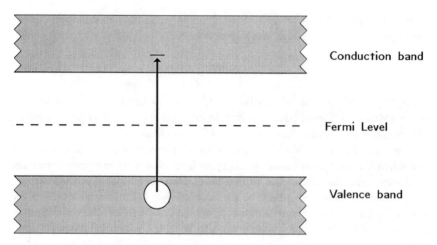

Figure 3.13. Position of the Fermi level in an intrinsic semiconductor.

Hence, the Fermi level of such an intrinsic semiconductor lies exactly midway between the top of the valence band and the bottom of the conduction band (Fig. 3.13). If we introduce donor or acceptor levels into the intrinsic semiconductor, the Fermi level will change. In an n-type semiconductor the level will usually lie between the donor levels and the conduction band and for a p-type semiconductor, it will be between the valence band and the acceptor levels as shown in Fig. 3.14.

Light of a given wavelength will be transmitted through a semiconductor of

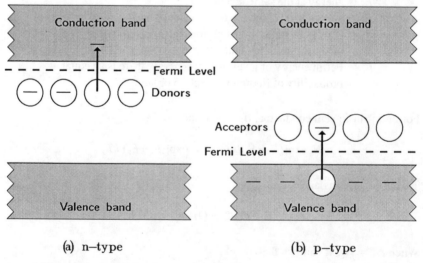

(a) n–type   (b) p–type

Figure 3.14. Position of the Fermi levels in doped (a) n-type and (b) p-type extrinsic semiconductors.

given thickness if the incident photons do not possess sufficient energy $h\nu$ to promote electrons from filled levels to empty ones. Otherwise, it will be absorbed to some extent, depending upon the density of states of the filled and empty levels. The fraction of incident energy absorbed is given by

$$(E_{abs}/E_{inc}) = 1 - e^{-\alpha t}$$

where t is the thickness of the semiconductor and the optical absorption coefficient, α, is a measure of the density of filled and empty states as a function of photon energy, and is a characteristic of the particular semiconductor. The energy above which the optical absorption increases rapidly is called the optical band gap, E_g, and corresponds to the separation between the top of the valence band and the bottom of the conduction band.

In some materials, an electron can be excited directly from the valence band to the conduction band without any change in its momentum. Then the optical gap is called a "direct" band gap, and the absorption coefficient increases as the square root of $(h\nu - E_g)$. In other materials, a more elaborate process is required to optically excite an electron (2). Then the optical gap is called an "indirect" band gap, and the absorption coefficient increases as the square of the energy difference $(h\nu - E_g)$ and is temperature dependent.

References

1. L. J. van der Pauw, *Philips Tech. Rev.*, **20**, 220 (1958).

2. C. Kittel, *Introduction to Solid State Physics*, 6th ed. John Wiley, New York, 1981.

Additional References

N. W. Ashcroft and N. D. Mermin, *Solid State Physics*, Chaps. 28 and 29. Saunders College, 1976.

A. K. Cheetham and P. Day (eds.) *Solid State Chemistry Techniques*. Clarendon Press, Oxford, 1987.

P. A. Cox, *The Electronic Structure and Chemistry of Solids* Chaps. 1 and 4. Oxford University Press, Oxford, 1987.

R. Hoffman, *Solids and Surfaces, A Chemist's View of Bonding in Extended Structures*, VCH, New York, 1988.

L. Solymar and D. Walsh, *Lectures on the Electrical Properties of Solids*, 4th ed., Chaps. 8 and 9. Oxford University Press, Oxford, 1979.

Problems

P3.1. A sample of silicon doped with antimony is an n-type semiconductor. Suppose a small amount of gallium is added to such a semiconductor. Describe how the conducting properties of the solid will change as a function of the amount of gallium added.

P3.2. Pure NiO is green and, when heated in oxygen at 800°C, turns black; the green form of the oxide is paramagnetic/antiferromagnetic and a relatively poor conductor of electricity. The black form has a much higher conductivity than the green form. Account for each of these observations.

P3.3. a. What is a degenerate semiconductor?

  b. How can you differentiate the electronic properties of a degenerate semiconductor from those of a metal?

  c. Indicate how you can increase the number of carriers in a single crystal of n-CdS.

  d. How could you experimentally determine the extent of solid solution between the compounds GaP–ZnS?

P3.4. TiO and VO show metallic behavior whereas MnO, CoO, and NiO do not. The conductivity of the latter three compounds can be increased by heating them in slightly oxidizing conditions. For MnO and CoO under more vigorous oxidizing conditions, new phases are formed.

  a. Describe the conduction process in TiO and VO.

  b. Describe the conduction process in NiO and $Ni_{1-x}O$.

  c. Indicate what phases are formed (i.e., structure types) when CoO and MnO are oxidized in oxygen at elevated temperatures.

P3.5. The electrical conductance of germanium is increased by a factor of 10^5 when a few parts per million of arsenic are added. Explain.

P3.6. a. How would you convert (chemically) an intrinsic silicon single crystal wafer to an n-type conductor? To a p-type conductor?

  b. Clearly show the Fermi levels as well as the $E_{v.b.}$ and $E_{c.b.}$ using simple band pictures to illustrate both of the above cases.

  c. How can silicon be made to behave like a degenerate semiconductor?

4

Magnetochemistry

All materials can be classified according to their behavior in a magnetic field. This is possible because a field is either present or can be induced in the materials, and this internal field interacts with the applied (or external) field. If the induced field opposes the external magnetic field, the material is said to show diamagnetic behavior. When the induced field aids the external field, the material is called paramagnetic. Hence, when a material is placed in an inhomogeneous magnetic field, it is either attracted toward a strong magnetic field (paramagnetic) or repelled from it (diamagnetic). An electron may be considered as a magnet formed by an electric charge spinning on its axis. Paramagnetism results from the presence of a permanent magnetic dipole, and is characteristic of atoms or molecules with unpaired electrons. In addition, an electron traveling in a closed path around the nucleus produces a magnetic moment. Filled electron shells are diamagnetic. The magnetic properties of an atom will result from the combined spin and orbital moments of its electrons.

The magnetic moment of an atom, ion, or molecule can be expressed in units of Bohr magnetons (BM or μ_B), defined by the expression

$$1 \mu_B = eh/4\pi mc = 0.927 \times 10^{-20} \, \text{erg/G}$$

where

$e =$ electronic charge
$m =$ electron mass
$h =$ Planck's constant
$c =$ speed of light

For a single electron, the magnetic moment μ_s is shown from quantum mechanics to be

$$\mu_s(\text{in BM}) = g\sqrt{s(s+1)}$$

where

$$s = \text{spin quantum number}$$
$$g = \text{gyromagnetic ratio}$$

The spin-only quantum number has a value of 1/2 and the gyromagnetic ratio is 2.00. If these values are substituted for s and g in the above equation, it gives a value of $\mu_s = 1.73$ for a single electron. For atoms or ions with more than one unpaired electron, the equation can be written as

$$\mu_s = g\sqrt{S(S+1)}$$

where S is the sum of the spin quantum numbers of the individual unpaired electrons. Thus high spin Mn^{2+} contains five unpaired electrons, $3d^5$, so that $S = 5/2$ and $\mu_s = 5.92$ BM.

In some materials, an orbital moment in addition to the spin moment contributes to the magnetic moment. If both the spin and orbital moments make their full contribution, then, for transition metals other than the lanthanides,

$$\mu_{S+L} = \sqrt{4S(S+1)+L(L+1)} = g\sqrt{S(S+1)} \text{ with } g \neq 2.00$$

where L is the orbital angular momentum quantum number of the ion. In the presence of a crystalline electric field, wave functions associated with different values of L are admixed. This results in the orbital angular momentum being partially or totally quenched, so that it becomes difficult to specify the relative spin and orbital contributions. Hence, one uses the form:

$$\mu_{S,L} = g\sqrt{S(S+1)}$$

with variable values for g.

From an experimental point of view, paramagnetic moments are not measured directly, but instead, the magnetic susceptibility is determined and then the magnetic moment can be calculated. If a substance is placed in a magnetic field, then

$$B = H + 4\pi I$$

where

$$H = \text{the magnitude of the magnetic field}$$
$$B = \text{the flux within the substance}$$
$$I = \text{the intensity of magnetization.}$$

The magnetic permeability is defined by the ratio B/H and is given by the equation

$$B/H = 1 + 4\pi(I/H)$$
$$= 1 + 4\pi k$$

where k is the magnetic susceptibility per unit volume (or the volume susceptibility). The permeability B/H can be considered as the ratio of the density of lines of force within the substance in the presence and absence of an applied field. The volume susceptibility of a vacuum is equal to zero, since in a vacuum $B/H = 1$.

In practice, magnetic susceptibilities are reported on the basis of weight rather than volume. Hence,

$$\chi = k/d$$

and

$$M \cdot \chi = \chi_M$$

where

$$d = \text{density in g/cm}^3$$
$$M = \text{molecular weight}$$
$$\chi = \text{gram susceptibility}$$
$$\chi_M = \text{molar susceptibility.}$$

The final value for χ_M is reported after a diamagnetic correction for closed electronic shells has been made. In some cases, the value for the diamagnetic correction can be determined experimentally and in others, the value is calculated from diamagnetic moments of the cations and anions listed in various tables (1). It may also be desirable to correct χ_M for the temperature-independent paramagnetism arising from a small polarization of any delocalized electrons, although this term is usually small and can be ignored.

Pierre Curie showed that, for an ideal paramagnet,

$$\chi_M^{\text{corr}} = C/T$$

where

$$C = \text{Curie constant}$$
$$T = \text{absolute temperature.}$$

It should be noted that the positive value of the susceptibility is consistent with the observation that a paramagnetic material is attracted to a strong magnetic field.

Langevin applied classical statistical mechanics to paramagnetism in gases and found that the Curie constant could be expressed as

$$C = N\mu^2/(3k)$$

where

N = Avogadro's number
k = Boltzmann's constant
μ = magnetic moment per molecule (in BM).

Hence,

$$\chi_M^{corr} = N\mu^2/(3kT) = C/T$$

at any given temperature, and

$$\mu = \sqrt{3k/N} \cdot \sqrt{\chi_M^{corr}T}$$

The value for $\sqrt{3k/N} = 2.84$ and thus

$$\mu = 2.84 \sqrt{\chi_M^{corr}T} = 2.84 \sqrt{C}$$

To calculate the magnetic moment of an ion, atom, or molecule, the weight susceptibility is measured and the corrected molar susceptibility is calculated. From this corrected susceptibility and the temperature, the magnetic moment is determined.

In practice, values for χ_M^{corr} are obtained as a function of temperature and a plot of $1/\chi_M^{corr}$ is made as a function of T. In some cases, a straight line is obtained similar to those shown in Fig. 4.1a. The slope is C^{-1} (the Curie constant) and the line intersects the origin. Other plots of $1/\chi_M^{corr}$ vs T resemble either line b or c where the T axis is cut either above or below 0 K.

Such a line requires a change of the Curie expression to

$$\chi_M^{corr} = C/(T - \theta)$$

where θ is the temperature at which the line b or c intersects the temperature axis and is known as the Weiss constant. The Weiss constant is related to the nature of the long-range interactions of the magnetic species present. If θ is negative, then the interactions are antiferromagnetic; a positive value of θ indicates long-range ferromagnetic interactions. These long-range interionic or intermolecular

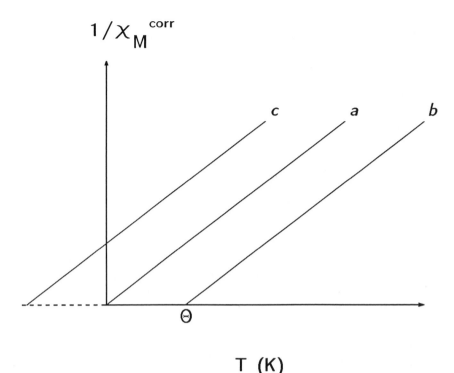

Figure 4.1. Variation of $1/\chi_M^{corr}$ with temperature (extrapolated from high temperatures) for (a) paramagnetic, (b) ferromagnetic, and (c) antiferromagnetic materials.

interactions require that the equation for determining the magnetic moment be rewritten as

$$\mu = 2.84 \sqrt{\chi_M^{corr}(T - \theta)} = 2.84 \sqrt{C}$$

This equation is valid at temperatures greater than $\approx 3\theta$.

It is now important to compare ferromagnetic and antiferromagnetic behavior with that of simple paramagnetism. If the susceptibility is plotted versus absolute temperature, three types of diagrams are possible for simple magnetic behavior, and these are shown in Fig. 4.2. Figure 4.2a represents a plot as given by the Curie law. Figure 4.2b represents the behavior of a ferromagnetic material. Above T_C, the Curie temperature, the material behaves according to the Curie–Weiss law; below this point, long-range ferromagnetic order of the ionic moments becomes established. The aligned parallel moments give rise to (spontaneous) magnetization even without any applied external field, and the apparent susceptibility (defined as M/H) becomes field dependent. The spontaneous magnetization (shown shaded in Fig. 4.2b) achieves its maximum (saturation) value at $T = 0$

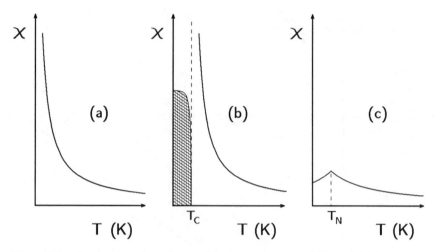

Figure 4.2. Temperature dependence of the magnetic susceptibility χ for (a) paramagnetic, (b) ferromagnetic, and (c) antiferromagnetic materials. Magnetic ordering occurs at the Curie and Néel temperatures, T_C and T_N, respectively. The shaded region below T_C represents spontaneous magnetization.

K. The saturation moment per ion is denoted by n, and is not the same as the thermal (paramagnetic) moment μ obtained at elevated temperatures.

Antiferromagnetic behavior is shown in Fig. 4.2c, where ideally a characteristic maximum in the susceptibility is observed at the Néel temperature, T_N. Above T_N, the sample behaves as a paramagnet with a negative Weiss constant; below T_N, long-range order of the ionic moments becomes established and the susceptibility decreases with decreasing temperature. A more complicated situation, called ferrimagnetism and not shown in Fig. 4.2, arises when the antiparallel moments are of unequal magnitude. At elevated temperatures, the magnetic susceptibility of such materials behaves like that of antiferromagnets. However, this susceptibility increases without limit as T_N is approached from above, and a net spontaneous magnetization, similar to that of a ferromagnet, signals the presence of long-range order below T_N.

Magnetic order is often discussed in terms of sublattices. In the paramagnetic state, individual ionic moments are disordered and the application of an external magnetic field merely biases their random orientations slightly to yield a small thermal average moment. This does not create a sublattice, nor does the ferromagnetic state, where all the ionic spins are aligned parallel to each other. But the ordered antiferromagnetic state is usually described as comprising two magnetic sublattices: one with spins pointed up and the other with spins down. Such sublattices arise when otherwise equivalent crystallographic sites are differentiated magnetically. Neutron diffraction is sensitive to electron spin as well as to atomic nuclei, and hence displays additional, non-nuclear peaks when magnetic sublattices are present. These may correspond to a doubling of the crystallographic unit cell.

Ferrimagnetism (unbalanced antiferromagnetism) can result from an ordering of vacancies onto just one antiferromagnetic sublattice. It is more common, however, when different magnetic ions occupy different crystallographic sites, as in the spinel structure. Then the magnetic sublattices coincide with crystallographic ones. But complications arise when antiferromagnetic interactions occur within each sublattice as well as between different sublattices. This situation gives rise to a canting of the spins away from a common axis, which is evidenced as a reduction in the observed magnetic moment compared to that calculated on the basis of coaxial spins.

References

1. P. W. Selwood, *Magnetochemistry*, Interscience Publishing, New York 1956.

Additional References

J. B. Goodenough, *Magnetism and the Chemical Bond*, John Wiley, New York, 1963.

D. C. Jiles, *Introduction to Magnetism and Magnetic Materials*, Chapman and Hall, London, 1990.

D. H. Martin, *Magnetism in Solids*, M.I.T. Press, Cambridge, MA, 1967.

Problems

P4.1. Indicate how you would distinguish between paramagnetic, ferromagnetic, and antiferromagnetic behavior of a solid using magnetic susceptibility measurements.

P4.2. Vanadium monoxide is Pauli paramagnetic and a good conductor of electricity, whereas nickel oxide is paramagnetic/antiferromagnetic and a poor conductor of electricity. Account for these observations.

P4.3. $ZnFe_2O_4$ is antiferromagnetic. Assuming that zinc atoms occupy the tetrahedral voids in the spinel structure, suggest a model that explains the antiferromagnetism in zinc ferrite.

P4.4. Each atom in an iron crystal contributes 2.22 μ_B on the average. However, the total number of Bohr magnetons per Fe_3O_4 formula in magnetite is calculated to be 14 μ_B. How can you justify this calculation? How do you account for the observed value of 4.08 μ_B per Fe_3O_4.

P4.5. Consider nickel ferrite having the formula $Fe[Ni_xFe_{2-x}]O_4$. What must the value of x be to give a net moment per formula weight of 2 BM?

P4.6. Make a sketch of an antiferromagnetic crystal having the CsCl structure. Assume that the two different kinds of atoms have antiparallel spin moments and indicate this by arrows in the two sublattice arrays. Indicate the antiferromagnetic unit cell in this crystal.

5

Phase Diagrams

To determine the relationship between the number of phases present in a system at thermal equilibrium and the number of components necessary to form these phases under given conditions of temperature, pressure, and composition, Gibbs developed the phase rule, which can be stated as

$$f = c - p + 2$$

where

$f =$ the number of variables necessary to describe the state of the system (the degrees of freedom)
$c =$ the number of components
$p =$ the number of phases.

In this treatment, other variables are disregarded. The quantity, f, deals with the number of variables that have to be specified to define the system completely, and usually includes the concentration of each phase. This number is called the variance or the number of degrees of freedom. In a system of two different chemical substances, if one has a concentration that gives a mole fraction of 0.3, the concentration of the other is fixed at 0.7 because the sum of the two must be 1.0. The variance in concentration is therefore 1, not 2.

The number of components c is the smallest number of independently variable chemical compounds that must be specified in order to describe the composition of each phase present in the system. For example, in the system liquid water, ice, and water vapor, there is only one chemical individual, namely water.

Consider a system in equilibrium consisting of p phases and c independent components. Since the sum of the mole fractions of the components is equal to 1.00, one component can be determined by difference if the others are known.

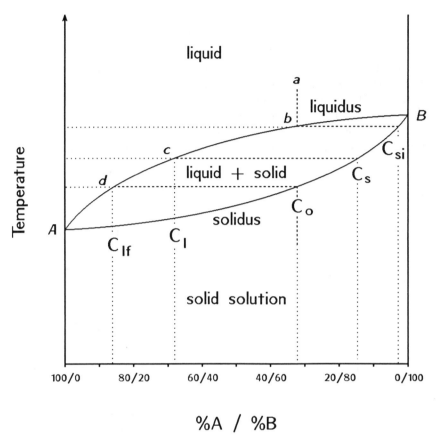

%A / %B

Figure 5.1. Diagram of a binary solid solution in which A and B are completely soluble in each other in all proportions in both the liquid and solid phases.

Hence, the composition of each phase can be defined by $(c - 1)$ concentrations. The compositions of all p phases can be defined by $p(c - 1)$. Because temperature and pressure can also vary, the total number of independent variables is $p(c - 1) + 2$. The chemical potential μ is the same in each phase α, β γ, etc. and hence for the component i

$$\mu_{i\alpha} = \mu_{i\beta} = \mu_{i\gamma} = \ldots$$

Hence, for a system of components there are $c(p - 1)$ independent equations and hence $c(p - 1)$ variables that are fixed when the system is in equilibrium. The number of independent variables = total number of variables − number of variables automatically fixed:

$$f = p(c - 1) + 2 - c(p - 1) = c - p + 2$$

The principal way in which phase diagrams are illustrated is as temperature–composition diagrams, and these will now be discussed briefly. Such diagrams define the phases that coexist, their compositional limits, and their thermal stabilities. Other important parameters such as pressure or activity are usually unspecified and the phase rule reduces to $f = c - p + 1$ when the pressure has been fixed. We will in this treatment of phase diagrams deal with a number of binary systems. For a more detailed treatment, the reader is referred to Refs. 1 and 2.

A. Binary Phase Diagrams

For binary systems, temperature is usually plotted as the ordinate and composition as the abscissa. A one-phase equilibrium has two degrees of freedom (since $c =$

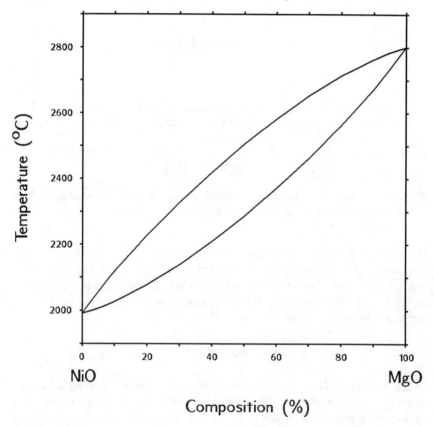

Figure 5.2. Phase diagram for the system MgO–NiO.

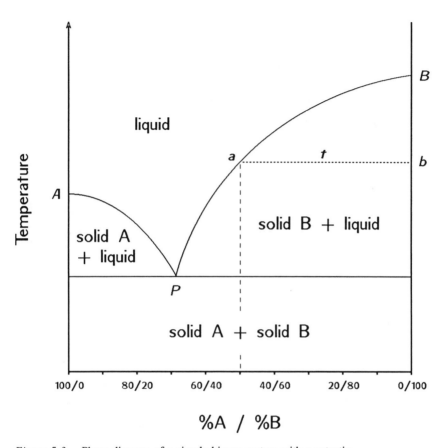

%A / %B

Figure 5.3. Phase diagram of a simple binary system with a eutectic.

2), namely, temperature and composition of the phase. If two phases are present, there is only one degree of freedom. Hence, the composition of both phases is determined if the temperature is indicated. A three-phase equilibrium has no degree of freedom, can be represented by a point, and is "invariant."

Figure 5.1 illustrates a binary solid solution diagram in which A and B are completely soluble in each other in all proportions in both the liquid and solid states. It can be seen that there is a single phase liquid and single phase solid region as well as a liquid plus solid two-phase region. The temperature–composition curves separate the single and two-phase regions. The temperature–composition curve that is in equilibrium with the solid is designated the liquidus curve. Similarly, the solidus curve is in equilibrium with liquid. If a horizontal, constant temperature line is drawn between the liquidus and solidus curves, then its ends represent the compositions of the two phases at the temperature indicated by the line. This type of diagram is commonly found for systems where the compounds

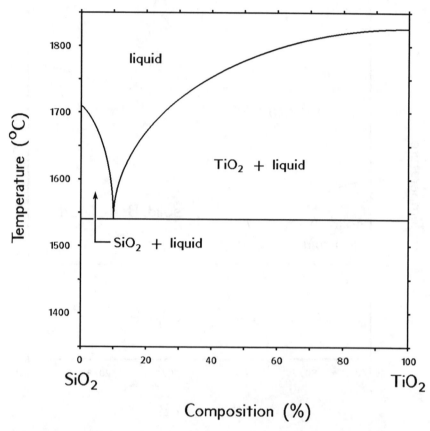

Figure 5.4. Phase diagram for the system SiO₂–TiO₂

of pure metals A and B form solid solutions or alloys of varying compositions. For example, in Fig. 5.1 the alloy composition C_0 at temperature a is a single phase liquid. On cooling to point b, solid begins to form and the composition is represented as C_{si}. Continued slow cooling to point c gives a solid composition of C_s and a liquid composition of C_ℓ. Further cooling to point d gives a final liquid composition of $C_{\ell f}$. Finally, at any temperature below point d, the material is completely solid with a composition C_0. The phase diagram for the MgO-NiO system, presented in Fig. 5.2, evidences such behavior.

Between the temperatures b and d, the alloy consists of two phases, one liquid and one solid phase. The relative amounts of the two phases present in a two-phase region can be determined by the so-called lever rule. Consider the dashed horizontal line at the temperature c in Fig. 5.1 sometimes called a tie line. If the weights of the phases at each end of the tie line are assumed to balance a mechanical lever whose fulcrum is at the alloy's composition, the condition of mechanical equilibrium requires that

$$W_\ell(C_\ell - C_0) = W_s(C_0 - C_s)$$

where W_ℓ and W_s are the weight fractions of the liquid and solid phases, respectively. The same rule also holds for two solid solutions in equilibrium with each other.

Figures 5.3 and 5.4 illustrate temperature–composition curves for simple binary systems with a eutectic. In Fig. 5.3, the two curves *AP* and *PB* are obtained if the freezing points of a series of liquid mixtures, varying in composition from one pure component A to the other B, are determined and plotted against the corresponding compositions of the liquid. The addition of component B lowers the freezing point of A as shown by curve *AP*; a similar lowering of the freezing point of B occurs on addition of A and is shown by line *PB*. When liquids rich in A are cooled between *A* and *P*, the solid A separates, and B separates when liquids rich in B, i.e., between *B* and *P,* are cooled. At the point *P* where the two curves meet, both solids A and B are in equilibrium with the liquid. Since

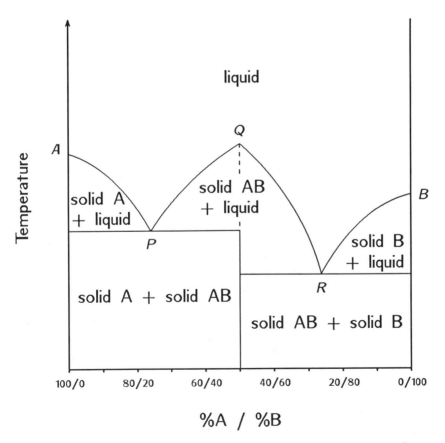

Figure 5.5. Phase diagram of a binary system in which the two components form a compound with a congruent melting point.

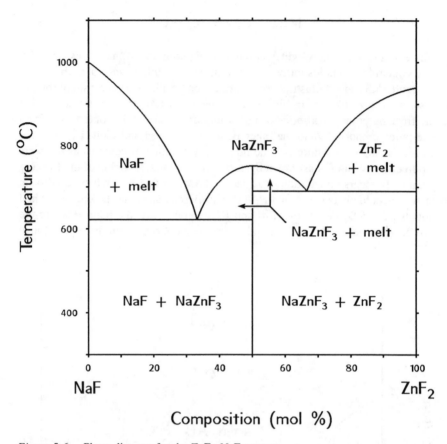

Figure 5.6. Phase diagram for the ZnF_2–NaF system

three phases coexist, the condensed system becomes invariant, i.e., there is only one temperature (at atmospheric pressure) where the liquid phase is in equilibrium with both solids. The point is P and is called the eutectic point (Greek: easily melting); it is the lowest temperature for the existence of a liquid phase.

At a temperature t, there is an equilibrium between the liquid of composition a and the pure solid at point b. Any point between a and b represents a mixture of liquid a and solid b in various proportions. The closer the point is to a, the less solid is present, whereas if it is nearer b, relatively more solid is found.

Figures 5.5 and 5.6 illustrate systems in which two components form a compound with a congruent melting point. In Fig. 5.5, the two curves AP and RB represent the compositions of liquid in equilibrium with solid A and solid B, respectively. There is a central portion PQR that rises to a maximum. This portion of the curve represents liquid systems in equilibrium with solid compound AB. The maximum Q of the curve occurs at the composition of the solid compound.

In Fig. 5.5, the compound contains equimolar amounts of A and B. Hence, the point Q is midway between pure A and pure B. At point Q, the composition of the solid and liquid phases is the same and the melting point of compound AB is said to be congruent.

Figure 5.5 shows two eutectic points, namely, P and R. Liquid compositions rich in A separate as solids A and AB on freezing. Similarly, solids B and AB separate for compositions rich in B.

The final illustrations of binary phase diagrams are given in Figs. 5.7 and 5.8. Sometimes a compound decomposes at a temperature below its melting point. In this case, the solid cannot be in equilibrium with a liquid having the same composition. Hence, no true melting point exists. Figure 5.7 illustrates the formation of a compound AB_2. At a temperature Q, which is below its hypothetical melting point, the compound dissociates into its constituents, i.e., liquid of

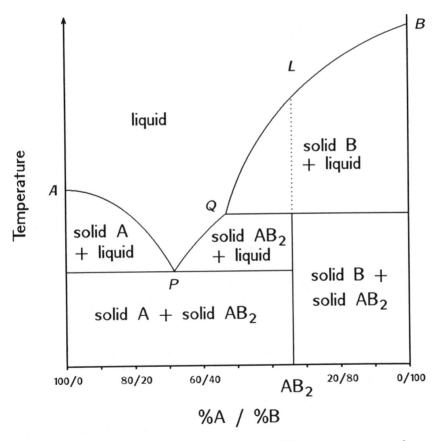

Figure 5.7. Phase diagram of a binary system in which the two components form a compound that melts incongruently.

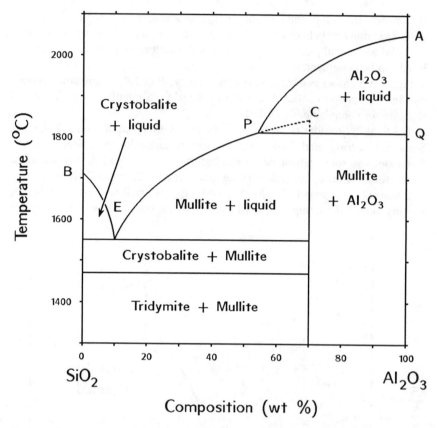

Figure 5.8. Phase diagram for the system Al₂O₃–SiO₂.

composition Q and solid B. Along curve PQ, the solid AB_2 separates from the liquid, but along BQ the solid B forms. The temperature Q is called the incongruent melting point of the compound. If a liquid, ℓ, is cooled, solid B will first separate and, when temperature Q is reached, solid AB_2 begins to form. The temperature Q will remain constant until the solid B has completely converted to solid AB_2 and the liquid phase disappears.

A well-known example of a system that has an incongruent melting point is the silica–alumina system, for which the phase diagram is shown in Fig. 5.8. Silica exists in two crystalline forms within the temperature range shown in Fig. 5.8, i.e., tridymite and cristobalite (which melts at 1710°C). Alumina crystallizes with the corundum structure and melts at 2050°C. The composition of the binary aluminum silicate (mullite) has the formula $3Al_2O_3 \cdot 2SiO_2$ (70 wt% Al_2O_3). The compound decomposes at 1810°C into Al_2O_3 and a liquid solution containing approximately 54 wt% of alumina.

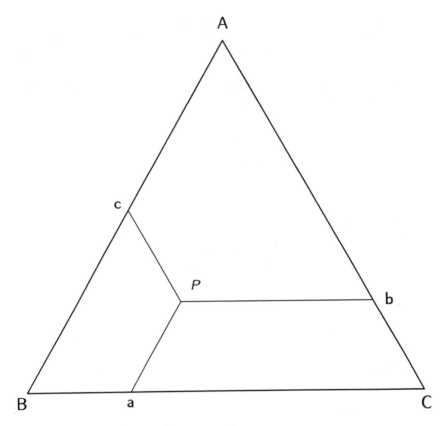

Figure 5.9. Phase diagram of a ternary system.

The single eutectic, E, is found between crystobalite and mullite. The areas below the freezing point curves indicate the equilibria between the phases shown. Below the curve EP at temperatures above the eutectic, the liquid solution is in equilibrium with mullite. At temperatures above P under curve PA, the liquid solution is in equilibrium with solid Al_2O_3. A phase transition occurs at P and this is called the peritectic point. Along the line PQ there are three phases in equilibrium, i.e., liquid of composition P, mullite, and Al_2O_3. The dotted curve PC represents a continuation of EP to indicate the maximum in the freezing point curve that would occur if the compound did not decompose at a lower temperature.

B. Ternary Phase Diagrams

The composition of a three-component system can be represented by an equilateral triangle. In such a triangle, in which all of the angles are 60°, the sum of the distances from any point within the triangle drawn parallel to the three sides is always equal to the length of the side of the triangle. Consider the equilateral triangle shown in Fig. 5.9 where there are three independent components A, B, C. The corners of the triangle represent the pure components A, B, and C. The distance relative to a triangle edge from point P to any given side measured parallel to either of the others gives the proportion of the component occupying the corner opposite the given side. Hence

$$Pa/BC = \text{fractional amount of A}$$
$$Pb/CA = \text{fractional amount of B}$$
$$Pc/AB = \text{fractional amount of C.}$$

If the length of a side of the triangle is unity, it is possible to express the amounts of the three components of a given system as fractions of the whole and hence to represent the composition of any system by a point in the diagram. A point on one of the sides of the triangle represents two components only, e.g., a point on side BC means that the concentration of A is zero, etc.

References

1. J. E. Ricci, *The Phase Rule and Heterogeneous Equilibrium*, Van Nostrand, New York, 1951.

2. S. Glasstone, *Textbook of Physical Chemistry*, 2nd ed. Van Nostrand, New York, 1946.

Additional References

A. Findlay, *The Phase Rule and Its Applications*, 9th ed. Dover, New York, 1951

Problems

P5.1. Can CsCl and NaCl form solid solutions? Explain. To what extent can NaCl form solid solutions with KCl?

P5.2. Figure P5.2 can be used to illustrate the principle of the triangular diagram. A, B, C, which are the corners of the triangle, represent the pure components A, B, C, respectively. Indicate clearly what Pa, Pb, and Pc represent in terms of A, B, and C.

P5.3. Explain why complete solubility can occur between two components of a substitutional solid solution but not for an interstitial solid solution.

P5.4. Interpret the phase diagram (Fig. P5.4) for CaO and ZrO_2 stating the phases at each area, line, and point.

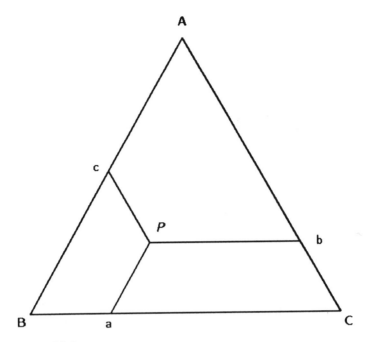

Figure P5.2

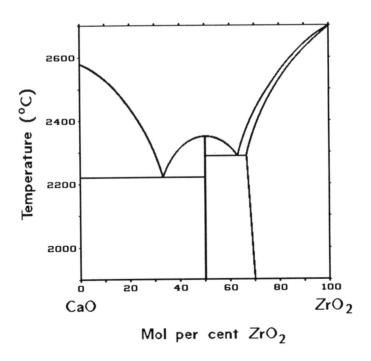

Figure P5.4

Part II
Transition Metal Oxides

Introduction

Oxides represent one of the most widely studied group of materials. The synthesis can in many cases be carried out under ambient conditions. Problems of chemical stability, purity, and homogeneity are less severe than for halides, pnictides, or chalcogenides.

In this treatment of transition metal oxides, a general discussion of many synthetic techniques is given. Methods for the growth of single crystals are also compared. Those procedures that the authors have used in their own studies are emphasized.

The chemistry of the oxides is discussed in terms of their composition, structure, and most interesting properties. The properties and structures are correlated by means of a one-electron energy model based on crystal and ligand field concepts that are familiar to chemists. A number of the oxides chosen have wide industrial usage.

6

Synthetic Techniques

Solid state chemistry research today is concerned with the effort to understand properties in terms of a particular composition and structure. However, there remains a need to distinguish between property studies carried out on ill-defined and well-defined materials. The procedure chosen for the preparation of a material should be given far more attention than it has received in the past. The quality of the starting materials, as well as the subtleties of the techniques used, will determine in large measure the quality of the final product. The complete control of preparative conditions is essential for an understanding of properties and their ability to be varied. In this discussion concerning the preparation of transition metal oxides, the writers will illustrate different aspects of problems encountered from their own research experience. It is hoped that these examples will demonstrate a number of different synthetic approaches as well as justify why a particular method was chosen. Unfortunately, much of what follows does not appear in textbooks, but the information is generally accepted among solid state chemists.

A. Preparation of Polycrystalline Solids

Direct reaction of the elements or single binary compounds is undoubtedly the most widely used method for the preparation of polycrystalline solids. For the reaction to proceed within a reasonable period of time, high mobility of the reactants is desirable as well as maximum contact surface between the reacting particles. The rate of diffusion and reactivity can be increased by raising the temperature or forming a more reactive precursor. In general, complete reaction can be achieved if the reaction temperature reaches two-thirds of the melting point of one of the solid reactants. To ensure maximum common surfaces of the different reactants, high surface area powders are usually used. These powders are completely mixed and can be compressed into tablets, thus increasing the contact surfaces of the reactants. For the preparation of oxides, which can be carried out in air, ceramic or platinum crucibles are used. It is necessary to start with the purest, assayed simple oxides, accurately weighed, intimately mixed, and reacted without the preferential loss of one reactant or the introduction of non-volatile impurities. To obtain a pure product, it is necessary to pulverize the reactants thoroughly and heat them a second or even a third time. This is particularly true of compounds prepared from refractory materials such as MgO.

The direct combination of two or more solids where a melt is not formed is generally time consuming. The process is usually a complex one that depends on the surface area as well as the defect structure. Unfortunately, the formation of a product tends to reduce the area of contact between reactants and also to reduce the rate of the reaction. The extent of product formation is influenced by the area of interfacial contact and the ease of diffusion through a product layer. A much smaller amount of product is obtained when particles of small surface area are in contact than when a "compact" of a fine powder, high surface area reaction mixture is reacted. The diffusion of reactants through a product layer depends on temperature, defect structure of product layers, grain boundary contacts, presence of impurities, and effectiveness of phase boundary contacts. Usually continued grinding between heat cycles facilitates chemical reactivity between solid phases. Not only is the surface area of the reaction mixture maintained, but also fresh surfaces of the reactants are brought into contact.

When only solid products form as a result of the reaction of two solids, studies of the reaction become experimentally very difficult to follow. Semi-quantitative measurement of the product with X-ray diffraction is probably the most widely used method for the determination of product formation. However, this method gives low accuracy for poorly crystallized substances or those that are formed with highly defective structures. For many single-phase products, one can nevertheless follow the cell dimensions as a function of composition. It is also possible to compare the observed changes of density with those calculated from X-ray data.

"Precursor" methods have been developed for the preparation of stoichiometric ferrites and chromites (1–3). These methods achieve excellent stoichiometry, low trace impurity content, and homogeneity approaching the maximum theoretically possible. The principal advantage of the "precursor" method is that the two metals are mixed on an atomic scale so that greater reactivity and more homogeneous products result than by direct combination of ground oxides. The procedures involve the crystallization from solution of precursor compounds that contain the metals in the desired atomic ratio. Any other element that may be present is volatilized at elevated temperatures. Decomposition of these crystalline precursors in air yields a mixture of metallic oxides which are finely divided and intimately mixed. The ignition of these compounds at relatively low temperatures ($<1200°C$) results in the formation of the desired product.

The precursor techniques used by Wickham et al. (1,3) to prepare ferrites with the composition MFe_2O_4 involved the thermal decomposition of oxalate (3) or pyridinate salts (1). The synthesis of ferrites from mixed oxalates yields homogeneous products in a short time. It was shown that iron(II) oxalate dihydrate $FeC_2O_4 \cdot 2H_2O$ can be coprecipitated from boiling solutions of salts with the corresponding oxalates of manganese(II), cobalt(II), nickel(II), or zinc(II) in the form of mixed crystals.

The products are finely crystallized, free-flowing powders that are handled easily and contain the iron and other metals mixed on an atomic scale. The mixed oxalates are decomposed to ferrites on heating in the presence of air:

$$MFe_2(C_2O_4)_3 \cdot 6H_2O + 2O_2 \rightarrow MFe_2O_4 + 6H_2O + 6CO_2$$

Only relatively low temperatures are required except in the case of manganese ferrite. However, values for the atomic ratio Fe/M deviating by less than 1% from the theoretical value of two are seldom obtained. This fact is a consequence of small differences in their solubility and the tendency of the oxalates to form supersaturated solutions. Atomic ratios Fe/M very close to the theoretical value of two were obtained by Wickham (1) by the thermal decomposition of crystal-lized salts $M_3Fe_6(CH_3CO_2)_{17}O_3OH \cdot 12C_5H_5N$. The products were ignited in air at temperatures chosen to give the correct oxygen content. The preparations of the compounds $M_3Fe_6(CH_3CO_2)_{17}O_3OH \cdot 12C_5H_5N$ reported by Wickham are given in Table 6.1.

The results of the chemical analyses are given in Tables 6.2 and 6.3 and indicate that the preparative methods described were reasonably successful for manganese, cobalt, and nickel ferrites. The lattice constants listed in Table 6.3 are in good agreement with the best values given in the literature. The saturation magnetic moments (n_∞), with the exception of that for manganese ferrite (1), are also in good agreement with accepted values. The value of n_∞ is less sensitive to small departures from stoichiometry than it is to variations in the thermal history of the sample that influence the distribution of cations between the tetrahedral and octahedral sites of the spinel structure.

Table 6.1. Preparation of the Compounds $M_3^{2+}Fe_6(CH_3CO_2)_{17}O_3(OH) \cdot 12C_5H_5N$

Starting materials compound	Quan-tity	Volume of pyridine solvent	Number of re-crystallizations	Yield (g)	%
$MgFe_3(AcO)_8(OH)_3 \cdot 10H_2O$	69	500	1	54.1	58
$Mg(AcO)_2 \cdot 4H_2O$	8.1				
$MnFe_3(AcO)_8(OH)_3 \cdot 5H_2O$	71	500	0	78	70
$Mn(AcO)_2 \cdot 4H_2O$	10.5		1	55	49
$CoFe_3(AcO)_8(OH)_3 \cdot 8H_2O$	44.7	500	1	28.5	45
$Co(AcO)_2 \cdot 4H_2O$	6.2				
$Co_4Fe_9(AcO)_{26}(OH)_9 \cdot 23H_2O$	80	500	2	24	25
$Ni_4Fe_9(AcO)_{26}(OH)_9 \cdot 23H_2O$	75	500	2	29	33

Table 6.2. *Chemical Analyses of Ferrites Prepared by Ignition of "Pyridinates"*

Ferrite	Weight iron (%) Observed	Theory	Weight M^{2+} (%) Observed	Theory	Mole Ratio Fe/M^{2+}	Ignition temperature (°C)	Active oxygen (eq/mol)
$MgFe_2O_4$	55.69 ± 0.03	55.84	12.41 ± 0.06	12.16	1.954 ± 0.012	1000	0
$MnFe_2O_4$	48.32 ± 0.02	48.43	23.68 ± 0.03	23.82	2.008 ± 0.005	1300[a]	0.034
$CoFe_2O_4$	47.55 ± 0.01	47.60	25.24 ± 0.08	25.12	1.990 ± 0.01	1000	0.004
$NiFe_2O_4$	47.54 ± 0.03	47.65	25.03 ± 0.01	25.05	1.997 ± 0.004	1000	0

[a] Quenched in nitrogen, contains 0.12 percent by weight excessive active oxygen.

Table 6.3. *Properties of Ferrites Prepared by Ignition of "Pyridinates"*

Ferrite	Cubic lattice constant a_o (Å)	Saturation magnetic moment[a] n_B (BM/molecule)
$MgFe_2O_4$	8.384 ± 0.001	1.37 ± 0.02
$MnFe_2O_4$	8.512 ± 0.001	4.5 ± 0.01
$CoFe_2O_4$	8.388 ± 0.001	3.57 ± 0.01[b]
$NiFe_2O_4$	8.338 ± 0.002	2.12 ± 0.01

[a] Measured at 4.2 K in a field of 10,000 oersteds.

[b] Cobalt ferrite is not saturated at 10,000 oersteds.

The "precursor" method has also been applied (2) to the preparation of the chromites, which have the formula MCr_2O_4 and possess the spinel structure. Previous methods employed for the synthesis of chromites consisted simply of preparing an intimate physical mixture of two appropriate oxides and heating this mixture to a sufficiently high temperature (1400–1700°C) to cause the two oxides to react to form the desired product. The method does not yield pure products easily because of the refractory nature of chromium(III) oxide and of many of the divalent-metal oxides involved. Tedious grinding procedures and heat treatments must be carried out, and extraneous impurities are usually introduced as a result of the grinding. Ignition at elevated temperatures occasionally results in the preferential loss of some of the divalent oxides (e.g., ZnO, CuO).

The precursor methods used by Whipple and Wold (2) to prepare a number of chromites are given in Table 6.4. As previously indicated, these precursors achieve excellent stoichiometry and homogeneity. The principal advantage of the "precursor" method is that the two metals are mixed on an atomic scale so that greater reactivity and more homogeneous products result than by heating a mixture of ball-milled (or mortar ground) oxides. The procedures involve the crystallization from solution of compounds containing chromium and another metal in the atomic ratio 2:1. Any other element that may be present is volatized at elevated temperatures. Decomposition of these crystalline precursors yields a mixture of

Table 6.4. Preparation of the Stoichiometric Chromites

Chromite	Precursor	Cr/M^{2+}	Yield of precursor (%)
$MgCr_2O_4$	$(NH_4)_2Mg(CrO_4)_2 \cdot 6H_2O$	$2.010^a \pm 0.002$	70
$NiCr_2O_4$	$(NH_4)_2Ni(CrO_4)_2 \cdot 6H_2O$	1.999 ± 0.002	75
$MnCr_2O_4$	$MnCr_2O_7 \cdot 4C_5H_5N$	$2.015^b \pm 0.002$	90
$CoCr_2O_4$	$CoCr_2O_7 \cdot 4C_5H_5N$	$2.012^a \pm 0.002$	90
$CuCr_2O_4$	$(NH_4)_2Cu(CrO_4)_2 \cdot 2NH_3$	2.000 ± 0.003	43
$ZnCr_2O_4$	$(NH_4)_2Zn(CrO_4)_2 \cdot 2NH_3$	1.995 ± 0.0006	77
$FeCr_2O_4$	$NH_4Fe(CrO_4)_2$	1.995 ± 0.003	80

^{a} Based on total Cr only.
^{b} Based on total Mn only.

the oxides MO and Cr_2O_3 which are finely divided and intimately mixed. The ignition of these compounds, at a relatively low temperature ($<1200°C$), results in the formation of the crystalline chromite. The characterization of chromites is presented and discussed later in the section on spinels.

For the preparation of a number of complex transition metal oxides, stabilization of high valence states requires the development of synthetic techniques that can be carried out at low temperatures. The complete solid state reaction of refractory oxides at a low temperature is a difficult problem. Direct combination of reactants to produce a mixed metal oxide requires high temperature heating with frequent regrinding. Reaction proceeds rapidly at first, but as the layer of product forms, diffusion paths become longer and the reaction slows down. Extremely small reactant particles of about several hundred angstroms in diameter can be prepared by freeze drying (4–6) or by coprecipitation (6, 7). As a result of smaller particle size of the reactants, the reactivity improves markedly. However, the diffusion paths still are rather large and hence "precursors" are the best method for achieving rapid reaction at low temperatures. The use of compound precursors, such as those used to synthesize ferrites and chromites (1–3), requires the stoichiometry of the precursor to correspond with that of the desired product. Where this is not possible, the use of solid solution precursors (8–10) can, in some cases, be as effective as compound precursors, and yet avoids the limitations of stoichiometry. This method was used, for example, by Horowitz and Longo (9) to study the phase relations in the manganese rich portions of the Ca–Mn–O system below 1000°C. Both $CaCO_3$ and $MnCO_3$ crystallize with the calcite structure, and hence a complete series of Ca–Mn carbonate solid solutions could be prepared. These precursors allowed the calcium to manganese ratio to be continuously varied so that the entire Ca–Mn–O phase diagram could be studied. Vidyasagar et al. (10) prepared several members of the system $Ca_{1-x}Fe_xCO_3$ where $x = 1/3, 2/5, 1/2, 2/3, 4/5,$ and $12/13$ by low-temperature decomposition of carbonate solid solutions. In addition, they prepared two members of the system $Ca_{1-x}Co_xCO_3$ with $x = 1/2$ and $1/3$.

Many solid state chemists have prepared complex oxides by codecomposition of the corresponding nitrates. Decomposition of nitrates results in the formation of reactive oxides, which can readily combine to form the desired product. In most cases, the nitrates are readily formed from the dissolution of the metals or carbonates in nitric acid. It should be noted that the decomposition of metal nitrates favors the formation of oxides with the metals in high oxidation states.

Another interesting synthetic technique, first described by Hilpert and Wille (11), appears to have been neglected in recent years. The authors prepared several ferrites by a solid state, double decomposition reaction which can be represented by the following equation:

$$M^{2+}Cl_2 + Na_2M_2{}^{3+}O_4 \xrightarrow{400-500°C} M^{2+}M_2{}^{3+}O_4 + 2NaCl$$

This procedure can be used advantageously when the more common methods become difficult, as in the preparation of mixed oxides containing very refractory components. Wickham et al. (1) successfully prepared Fe_3O_4, $FeCr_2O_4$, $FeTiO_3$, and $FeAl_2O_4$ by this procedure.

Brixner (12–14) used a flux reaction for the preparation of a variety of ternary oxides. The products are usually obtained in the form of small, well-defined, single crystals. In this technique, a salt melt ($CaCl_2$ or $BaCl_2$) serves both as a flux and a reactant. Brixner (14) reported the preparation of $BaFe_{12}O_{19}$ by this method. The phase separated as transparent red crystals and the reaction can be represented by the following equation:

$$BaCl_2 + 6Fe_2O_3 + H_2O \rightarrow BaFe_{12}O_{19} + 2HCl$$

This reaction occurs at 1300 K and it is essential that $BaCl_2$ hydrolyzes to BaO. Bichowski and Rossini (15) indicated that the ΔH for the hydrolysis of $BaCl_2$, i.e.,

$$BaCl_2 + H_2O \rightarrow BaO + 2HCl$$

is $+78$ kcal/mol. This heat of reaction must be supplied by the formation of $BaFe_{12}O_{19}$. The conversion of $BaCl_2$ to BaO is a slow reaction and hence, the product $BaFe_{12}O_{19}$ will form as transparent red crystals up to a few millimeters in length. It has been observed (12) that the most stable composition between the two constituent oxides will be the only composition if there is a significant difference between the heats of formation of competing phases. By the flux reaction technique, Brixner was able to incorporate a considerable quantity of chlorine with the formation of the halide phosphates and vanadates of strontium

Table 6.5. Reaction Products from $CaCl_2$ with Various Oxides

Oxide reacted	Product and comment
Fe_2O_3	$CaFeO_4$ in red transparent fibrous form
Al_2O_3	$Ca_3Al_{10}O_{18}$ as transparent hexagonal plates
Cr_2O_3	$CaCrO_4$ in dendritic form
SiO_2	Ca_2SiO_4 fine acicular particles, surface area 120 m^2/g
P_2O_5	Ca_2PO_4Cl (chlorspodiosite) as flat transparent platelets

$Sr_5(PO_4)_3Cl$ and $Sr_2(VO_4)Cl$ (13). Other examples of reaction products from $CaCl_2$ and $BaCl_2$ with various oxides are given in Tables 6.5 and 6.6.

Various synthetic techniques have been developed for the known oxide systems. It is not possible to outline all of the approaches; however, the choice of atmosphere under which the preparation is being carried out will determine the oxidation state of the constituent metallic species in the final product. In general, oxides require successive heat treatments with intimate grinding of the products after each heating. High surface area products >100 m^2/g usually are made by low temperature decomposition of an appropriate precursor.

Table 6.6. Reaction Products from $BaCl_2$ and Various Oxides

Oxide reacted	Product and comment
Fe_2O_3	$BaFe_{12}O_{19}$ red transparent ferromagnetic flakes
WO_3	$BaWO_4$ transparent slightly acicular rectangular prisms
SiO_2	$BaSi_2O_5$ fine acicular particles
PbO	$BaPbO_3$ brown crystals
TiO_2	$BaTi_3O_7$ transparent flat crystals

a. Sol Gel Synthesis of Transition Metal Oxides

Sol gel methods have been widely used for the synthesis of simple as well as complex metal oxides. Widespread interest in this method has developed since 1969 (16). The reactions take place in solution and involve the conversion of a molecular precursor to an oxide network by forming intermediate inorganic polymers. Most of the literature is concerned with the preparation of silicates, Al_2O_3, TiO_2, or ZrO_2. However, in 1986, Livage (17) applied the sol gel method to the synthesis of transition metal oxides. Livage et al. (18) published an excellent review of the sol gel process as it can be applied to the synthesis of transition metal oxides.

The sol gel method involves both hydroxylation and condensation of molecular precursors. The precursor may be either an aqueous solution of an inorganic salt or an organometallic compound. Aqueous solutions of transition metal salts can involve numerous molecular species whose complexity is affected by the oxidation state, pH, and concentration. In a review article published in 1988 (18), Livage et al. surveyed the preparation of many transition metal oxide systems, but in this treatment only TiO_2 gels and hydrous ferric oxide will be discussed.

TiO_2 gels can readily be prepared by adding a weak base [e.g., Na_2CO_3 or $(NH_4)_2CO_3$] to a solution of sodium titanate dissolved in concentrated hydrochloric acid (19–21). Sols can also be obtained by the addition of $TiCl_4$ or $TiO(NO_3)_2$ to acidic aqueous solution (22, 23). The TiO_2 particles crystallize with either the anatase or rutile structure depending on the pH and the nature of the other ions present in solution. Recent studies have started with metallorganic precursors, e.g., titanium alkoxides $Ti(OR)_4$ (R = Et, n−Bu, n−Pr, i−Pr, s−Bu). Hydrochloric or nitric acid is used as a catalyst for gel formation (24,25). An excellent review of gel synthesis from inorganic precursors was published in 1986 by Woodhead (26).

The hydrolysis and condensation of metal alkoxides involve two steps. Initially, there is a partial hydrolysis of the alkoxide, which results in the formation of an active OH group that reacts further to form polymeric species and finally a gel. This can be represented by

$$Ti(OR)_4 + H_2O \rightleftharpoons Ti(OH)(OR)_3 + ROH$$
$$\equiv Ti - OR + HO - Ti \equiv \rightleftharpoons \equiv Ti - O - Ti \equiv + ROH$$

The process usually involves the slow addition of a water–alcohol solution to an alcoholic solution of the alkoxide. The gelation process is influenced by the presence of acid or base (27). An organic acid can be added to an alcoholic solution of the alkoxide. The water formed as a product can then hydrolyze the alkoxide and form a gel. Sintering of the gel results in the formation of high surface area TiO_2. Furthermore, a hydrated salt of a second metal can be dissolved in the alcoholic solution of the titanium alkoxide. The resulting products can be decomposed to give ternary oxides such as $BaTiO_3$. The gel route from metal–organic precursors provides the possibility of synthesizing many interesting and potentially useful materials. The aggregation of the colloidal particles is dependent on the sign and magnitude of surface charges (28) with respect to the pH of the solution.

Finally, particles of γ−Fe_2O_3 10 nm in diameter can be prepared by raising the pH of an aqueous solution containing both Fe^{2+} and Fe^{3+}. Stable sols are then formed by peptizing the flocculate in either a basic or acid medium (29).

b. Chimie Douce

Chimie douce or "soft chemistry" was first introduced to the scientific community by Rouxel and Livage in the mid-1970s (30). A variety of chemical reactions can be included as part of this novel approach to the synthesis of solids. They include cationic exchange, dehydration, dehydroxylation, hydrolysis, redox, intercalation, deintercalation, etc. Many of the products synthesized are thermodynamically metastable, although they can be kinetically stable at high temperatures. Their structures are novel and are related to the corresponding precursor structure. Many papers have been published recently that describe the versatility of this method. Among them are two papers by Rouxel (31) and Livage (32) describing the use of chimie douce to prepare low-dimensional solids and sol gels. A special issue of the *Revue de Chimie Minéral* is devoted to chimie douce (33).

The various processes that are considered to be "soft chemistry" generate new metastable phases. They utilize conditions that are gentle and avoid any structural transformation from the metastable phase to a stable one. The reaction used does not possess sufficient energy to produce a stable state. Usually the precursors are poorly crystallized, high surface area particles necessary to give increased reactivity. Whereas this may be considered as a restriction of "soft chemistry" methods, several examples will be given that indicate the potential usefulness of this technique.

A new form of TiO_2, namely $TiO_2(B)$, was synthesized by the hydrolysis of $K_2Ti_4O_9$ in dilute nitric acid to give $H_2Ti_4O_9 \cdot H_2O$ (34). This product was filtered, vacuum dried and then heated to 500°C to form the desired $TiO_2(B)$. The reversibility of the hydrolysis product reformulated as $H_3OTi_4O_8(OH)$ supports the belief that during hydrolysis the basic framework of the Ti_4O_9 sheet is not destroyed. Apparently, one of the nine oxygens is not shared between TiO_6 octahedra and can be readily hydroxylated. $TiO_2(B)$ is less compact than the other forms of TiO_2. The volume of a TiO_2 unit in $TiO_2(B)$ is 35.27 $Å^3$ whereas that for rutile is 31.12 $Å^3$, for anatase 34.01 $Å^3$ and for brookite 32.20 $Å^3$. The structure is monoclinic $a = 12.163(5)$ Å, $b = 3.735(2)$ Å, $c = 6.513(2)$ Å and $\beta = 107.29(5)°$. It is also possible to insert up to 0.85 Li in $TiO_2(B)$ either electrochemically or with BuLi. $TiO_2(B)$ is isotypic with $VO_2(B)$, which has been prepared by the low temperature reduction of V_2O_5 with hydrogen (35).

$Ni(OH)_2$ is usually precipitated from an aqueous nickel salt solution. The particles are small (36) and have a high surface area. However, the poor X-ray diffraction patterns of these products make the interpretation of the structure of this product difficult. By the use of chimie douce methods, $NaNiO_2$ (synthesized at high temperature) can be converted to a new variety of crystallized divalent nickel hydroxide (36). The methods employed at room temperature were successive exchange and reduction reactions of the precursor $NaNiO_2$ by the presence of intercalated water molecules in the van der Waals gap and by hydroxide layers stacking. However, the NiO_2 slabs of the $NaNiO_2$ precursor are maintained. The methods of chimie douce have been applied to the synthesis of new WO_3–MoO_3 oxides and a pyrochlore WO_3, as well as the preparation of new metastable sulfides Mo_6S_8, Mo_9S_{11}, $Mo_{15}Se_{19}$. From these binary phases, numerous new ternary metastable chalcogenides have been obtained by intercalation (37).

B. Sintering and Crystal Growth

When a solid is heated, it may be observed to undergo physical and/or chemical changes. These include sintering, melting, and thermal decomposition. The process of sintering probably results from crystal growth at the contact area between adjacent crystallites. This process proceeds at the expense of one of them. As a result, the crystallites become bonded together and the average size is observed to increase.

At elevated temperatures, the ions become mobile and melting occurs. The ordered lattice array is replaced by the short-range order of the liquid state. Crystallization may proceed by several different paths:

1. Vapor $\rightarrow$ solid (condensation)
2. Solution $\rightarrow$ solid (precipitation)
3. Melt $\rightarrow$ solid (freezing)
4. Solid A $\rightarrow$ Solid B (transformation).

All of these processes have in common the condition that for crystallization to proceed the final crystals must have lower free energy than the initial state of the system. The process of crystal formation involves two steps: first, the formation of a new nucleus and second, the growth of this nucleus to form a particle of appreciable size. Crystals may contain many imperfections or "defects" other than impurity atoms and these defects are most important in determining the properties of the crystal. In addition to defects, distortion of the lattice through departures from perfect alignment of repetitive units can exist. This type of imperfection is called a "dislocation." Hence, dislocations are regions within the solid at which the regularly repeating lattice array shows a discontinuity or a distortion from the ideal alignment of units within the crystal. Many important properties of crystals are dependent on the number and type of dislocations present; for example, a dislocation represents a region of weakness.

C. The Growth of Oxide Single Crystals

a. Growth from the Melt

There are five general methods for the growth of crystals from the melt that will be discussed in this section and have been in use for many years. The flame fusion method was invented by Verneuil (38) for the production of synthetic ruby crystals. The apparatus is shown in Fig. 6.1. The starting material is a mixture of $NH_4Al(SO_4)_2 \cdot 12H_2O$ and Cr_2O_3, which is kept in a hopper with a fine mesh screen at its base. Vibration of the sieve causes the powder to be fed into an

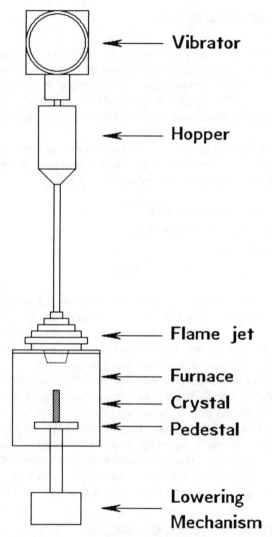

Vibrator

Hopper

Flame jet

Furnace

Crystal

Pedestal

Lowering
Mechanism

Figure 6.1. Schematic representation of the Verneuil apparatus

oxygen stream that flows through the central tube of an oxyhydrogen burner. Hydrogen is also supplied to the flame, which is contained in a cylindrical ceramic muffle. At the base of the muffle is a pedestal whose height can be adjusted. The powder that falls through the flame builds up a sintered cone on top of the pedestal. The tip of this cone is melted by increasing the flow of hydrogen to the flame. As more powder falls on the cone, the alumina pedestal is lowered at a uniform slow rate. The lowering rate is adjusted so that the surface of the molten cap remains at a fixed position in the flame. If the original cone is sufficiently narrow so that only one nucleus forms, the resulting solid is a single crystal. One problem with this method is the cracking of many crystals because of steep temperature gradients from the top to the bottom of the boule during crystal growth. A further problem is the reoxidation that occurs subsequent to crystallization as the surface of the boule cools. The boules are usually carefully annealed to remove the internal stresses induced during the growth process.

The Stöber method of crystal growth from the melt (39) places a crucible loaded with material in a furnace with a temperature gradient so that the top of the crucible is hotter than the bottom. Usually a crucible with a conical tip is used. Initially the temperature of the entire crucible is above that of the melting point of the crystal to be grown. The power to the furnace is reduced and, if there is a single nucleus formed and no complications, a single crystal will be grown. Reduction of the power to the furnace slowly allows the crystal to be annealed in situ.

The Bridgman–Stockbarger method (40) utilizes a furnace with a steep temperature gradient situated approximately at its center. The top half of the furnace is the higher temperature zone. The crucible is supported from below and has a controlled heat leak to the bottom. The crucibles, in general, have conical tips and are moved with respect to the furnace. The crucible is first placed in the upper part of the furnace and remains there until all of the material is molten. It is then drawn through the temperature gradient into the lower portion of the furnace, which is at a temperature below the melting point of the crystal being grown. After the boule has completely solidified, it can be annealed in the lower portion of the furnace by slowly reducing the power.

The Czochralski method (41) makes use of a furnace without a temperature gradient. The contents of the crucible are melted and a rod with a seed crystal attached to one end is lowered into the melt. The rod is cooled, creating the necessary temperature gradient. The rod is rotated and slowly withdrawn from the melt and the crystal is pulled out as the rod moves. This method is undoubtedly one of the most versatile methods for crystal growth from melts despite the difficulties involved.

Skull melting is another high-temperature method for the growth of oxide crystals. A power supply of 50 kW at 3 MHz has been used (42) to produce radio frequency power that is transferred to a power coil. The coil is wrapped about a skull crucible, which can be water cooled. Provision is also made for adjusting the ambient atmosphere to obtain a single phase of the proper stoichiometry. A graphite susceptor ring is inserted with the charge into the skull container. The graphite ring couples the charge to the power source so that heating can be achieved even if the conductivity of the charge is low. The process is crucibleless since the molten charge is isolated from the copper container by a thin layer of sintered material, which is next to the water-cooled skull. The graphite susceptor is burnt off as CO or CO_2 during the growth process. To achieve stabilization of mixed oxidation states (e.g., Fe_3O_4) the chamber above the container is closed, evacuated, and then can be filled with CO_2 (or CO/CO_2 mixtures). The boule is kept under a controlled atmosphere and the crucible can be lowered out of the stationary power coil. Relatively large crystals (few centimeters) of numerous oxides have been grown by this method.

b. High-Temperature Solution Growth

In this technique the constituents of the crystals to be grown are dissolved in a suitable solvent and crystallization occurs as the solution becomes critically supersaturated. Supersaturation can be achieved by evaporation of the solvent, cooling the solution, or transporting the solute from a hot to a cooler zone. The growth of crystals from a solvent is of particular value for compounds that melt incongruently since crystal growth occurs at a lower temperature than that required for growth from the melt. Unfortunately, crystal growth from solution usually results in the incorporation of solvent into the crystal.

Liquid phase epitaxy (LPE) is an important process in which a thin layer of crystalline material is deposited onto a substrate of similar composition or surface structure. This technique is used specifically for the deposition of thin films and requires that there is a match between the lattice parameters of the film and the substrate.

c. Flux Growth

One of the most widely used techniques classified as a high-temperature solution method is flux growth. Crystals of ceramics, ferrites, and other oxides have been grown by the slow cooling of a solution in a molten flux. For examples of this technique the reader is referred to the work of Remeika (43,44), who has grown barium titanate crystals from molten KCl and some ferrite crystals from molten PbO. One serious disadvantage of growing from such solutions is that the crystals usually contain traces of solvent.

d. Chemical Vapor Transport

Chemical-transport reactions have been used (45) to prepare single crystals of triiron tetroxide (magnetite) and other ferrites, which comprise the majority of all known ferrimagnetic materials. The work of Darken and Gurry (46) suggested that stoichiometric magnetite powder could be prepared by heating iron(III) oxide in an atmosphere of CO and CO_2, and pure Fe_3O_4 starting material for transport was prepared by this method (47).

The procedure that Hauptman (45) used for the growth of ferrite crystals can be summarized as follows: the powdered charge material is introduced into a silica tube, which is then evacuated. The transport agent is introduced and the tube sealed off. The tube is then placed in a two-zone transport furnace that has a temperature difference between the zones. The powdered charge material reacts with the transport agent to form a more volatile compound. The vapor diffuses along the tube to a region of lower temperature, where some of the vapor undergoes the reverse reaction. The starting compound is reformed and the transport agent is liberated. The latter then is free to react once again with the charge. Under the proper conditions, the compound is deposited as crystals. The transport of Fe_3O_4 using HCl as the transport agent occurs by the reversible reaction:

$$Fe_3O_4 + 8HCl \rightleftharpoons FeCl_2 + 2FeCl_3 + 4H_2O$$

This reaction is endothermic. Hence, the equilibrium shifts to the left at lower temperatures, and Fe_3O_4 is deposited in the cooler zone. For exothermic reactions, the desired product would be deposited in the hotter zone.

The same transport procedure has been used to prepare crystals of other metal oxides. Gray et al. (48) used chemical vapor transport to prepare single crystals of V_2O_3. Single crystals of V_2O_3 grown with HCl as the transport agent exhibited a first-order electrical transition from a metal to a semiconductor at 158 K on cooling, in agreement with the results first reported by Morin (49). However, single crystals prepared by chemical vapor transport using $TeCl_4$ as a transport agent remained metallic to 96 K on cooling. The suppression of the semiconducting phase was reported by Pouchard and Launay (50) and was attributed to the oxidation of V(III) to V(IV) by Cl_2. It was demonstrated (48) that such crystals contained V_3O_5. The procedure of chemical vapor transport can produce products that contain controlled amounts of several oxidation states of the transition metal. An excellent review of chemical vapor transport was written by Schäfer (51).

e. Hydrothermal Synthesis

A number of metal oxides and other compounds practically insoluble in water up to its boiling point show an appreciable solubility when the temperature and pressure are increased above 100°C and 1 atmosphere, respectively. These materials can be grown by the hydrothermal method. Oxides are usually grown from alkaline solutions (52) and metals from acid solutions (53). An excellent review of the role of hydrothermal synthesis in preparative chemistry was published by Rabenau (54). The method was introduced into modern solid state chemistry as a result of the method's success in the growth of quartz oscillator crystals. Hydrothermal growth can be considered as a special case of chemical vapor transport synthesis. This process combines a chemical reaction with a transport process, e.g., in the case of SiO_2 under hydrothermal conditions

$$SiO_2(s) + nH_2O(g) \rightleftharpoons (SiO_2 \cdot nH_2O)(g)$$

When the compounds are sparingly soluble with high melting points, a "mineralizer" is added to increase the solubility of the phase being grown.

Hydrothermal synthesis offers a number of advantages over other methods of crystal growth. Compounds containing elements in unusual oxidation states can be obtained in the closed systems utilized by hydrothermal synthesis. In this limited treatment, only a few examples can be given. Ferromagnetic chromium oxide, CrO_2, which is utilized in magnetic tapes, is produced by the oxidation of Cr_2O_3 with an excess of CrO_3:

$$Cr_2O_3 + CrO_3 \xrightarrow[H_2O]{350°C, 440\,bar} 3\,CrO_2$$

$$CrO_3 \xrightarrow[H_2O]{350°C, 440\,bar} CrO_2 + \tfrac{1}{2}O_2$$

The decomposition of the excess CrO_3 results in a build-up in the oxygen pressure, which stabilizes the product CrO_2. The product is produced in one step and is quite pure and uniform (55).

In a number of materials, the stabilization of low oxidation states can be achieved by reaction of the metal with water, e.g., vanadium to form $VO(OH)_2$ (56) and iron to form magnetite (57). A variation of this method has been used to transform a metal oxide with the corresponding metal into oxyfluorides represented by

$$nM + (m-1)M_nO_m \xrightarrow[pressure]{HF} mM_nO_{m-x}F_x$$

The oxyfluorides $V_2O_{5-x}F_x$ ($x \leq 0.025$), $VO_{2-x}F_x$ ($x \leq 0.2$) (58), as well as $WO_{3-x}F_x$ ($0.17 \leq x \leq 0.66$) and $MoO_{3-x}F_x$ ($0.74 \leq x \leq 0.97$) crystallize with the cubic ReO_3 structure (59,60).

In addition to the growth of quartz crystals and chromium(IV) oxide for magnetic tapes, hydrothermal crystal growth has been used to produce artificial gems. Synthetic emeralds (61) are produced by growth at 500–600°C and 1 kbar. Hydrothermal synthesis has also been used for the production of massive quantities of synthetic zeolites, which are used as molecular sieves. Their crystallization is carried out above 100°C in a closed hydrothermal system. This is especially true for the silicate-rich variants, e.g., mordenite (62).

f. High-Pressure Synthesis

The growth of crystals in the kilobar region requires high pressures, which can be generated by the action of a hydraulic ram operating on a piston. By a combination of several rams or pistons, it is now possible to maintain pressures of 200 kbar at temperatures in excess of 2000°C. An excellent review of high-pressure techniques is given in the book by Paul and Warschauer (63). Though published in 1963, the techniques described are still being used today with little modification. A simple arrangement for crystallization under high pressure is that where the sample is placed between a piston and a fixed closure within a cylindrical jacket. The maximum pressure that can be achieved is limited by the bursting pressure of the containing cylinder. In the opposed anvil arrangement, the cylinder is eliminated and the specimen is held in a small central region between two anvils which are tapered away from this region at about 10°. A layer of insulation can be inserted into a recess in the anvil and the pressure limitation is then set by the closeness of approach of the opposing anvils. This is governed by the behavior of the ring of pyrophyllite which behaves as a fluid under pressure and forms a gasket enclosing the specimen. A longer compression stroke is possible when curved anvils are employed, and the so called "belt" design is now widely used.

Large specimens can be prepared with multiple piston arrangements of which the tetrahedral-anvil apparatus is most common. Four anvils are located at the ends of four rams and enclose a tetrahedron of pyrophyllite. The later acts as a pressure transmitting medium and is bored out to contain the sample and heating element. Normally, pressure is applied to only one ram and the other three are mounted in a support ring. Non-uniform stresses must be avoided.

g. Electrolytic Reduction of Fused Salts

Andreiux (64,65) showed that it was possible to obtain a number of transition metal oxide single crystals by the electrolysis of molten salts containing mixtures of the appropriate oxides. Andreiux and Bozon (66,67) were able to prepare a number of vanadium spinels by electrolyzing melts of sodium tetraborate and sodium fluoride in which were dissolved the appropriate mixtures of transition metal oxides. The products were shown to have the composition MV_2O_4 (M = Fe, Mn, Co, Zn, Mg).

The reduction of TiO_2 or $CaTiO_3$ dissolved in a calcium chloride melt has been reported by Bertaut and Blum (68,69) and Bright et al. (70). The electrolysis was carried out by Bright at 850°C for 10 min using a current of 14 Å. Black, lustrous crystals formed at the cathode, which had the composition $CaTi_2O_4$. It has not been possible to synthesize this compound except by electrolysis.

Dodero and Déportes (71), have prepared a number of compositions of $NaMO_2$ (M = Fe, Co, Ni) by the electrolysis of sodium hydroxide melts contained in alumina crucibles. Electrodes of iron, cobalt, or nickel were used, depending on the desired composition of the final product.

Crystals of tungsten and molybdenum oxide "bronzes" have been grown by electrolytic reduction of tungstate or molybdate melts. Extensive reviews of the preparation and properties of the bronzes studied through 1980 are given in three review papers (72–74).

These are only representative examples of unusual transition metal oxides, which can be prepared (usually as single crystals) by electrolysis of fused salts. The methods of preparation of transition metal oxides that have been discussed here are representative techniques based on the authors' experiences. In all cases, purity of starting materials and care to avoid contamination during the synthesis are essential if meaningful characterization of the product is to be obtained. Often, X-rays of the products are not sufficient to ensure completeness of reaction. They must usually be supplemented with other characterization techniques. Finally, the synthesis of materials is often undervalued as an important component in valid scientific research.

References

1. D. G. Wickham, E. R. Whipple, and E. G. Larson, *J. Inorg. Nucl. Chem.*, **14**, 217 (1960).

2. E. Whipple and A. Wold, *J. Inorg. Nucl. Chem.*, **24**, 23 (1962).

3. D. G. Wickham, *Inorganic Syntheses* Vol. IX. McGraw Hill, New York, 1967, p. 152.

4. F. J. Schnettler, F. R. Monforte, and W. E. Rhodes, in *Science of Ceramics*, Vol. 4, G. H. Stewart, ed. *The British Ceramic Society*, 1967 p. 179.

5. Y. S. Kim and F. R. Monforte, *Am. Ceram. Soc. Bull.*, **50**, 532 (1971).

6. T. Sato, C. Curoda, and M. Saito, Ferrites: *Proc. Int. Conf. Jpn.* **72** (1970).

7. A. L. Stuijts, in *Science of Ceramic* Vol. 5, C. Brosset and E. Knapp, eds. *Swedish Institute of Silicate Research*, 1970, p.335.

8. L. R. Clavenna, J. M. Longo, and H. S. Horowitz, U. S. Patent 4,060,500 to Exxon Research and Engineering Co. (November 29, 1977).

9. H. S. Horowitz and J. M. Longo, *Mat. Res. Bull.*, **13**, 1359 (1978).

10. K. Vidyasagar, J. Gopalakrishnan, and C.N.R. Rao, *Inorgan. Chem.*, **23**, 1206 (1984).

11. S. Halpert and A. Wille, *Z. Phys. Chem.*, **18B**, 291 (1932).

12. L. H. Brixner and K. Babcock, *Mat. Res. Bull.*, **3**, 817 (1968).

13. L. H. Brixner, *Inorganic Syntheses*, Vol. XIV. McGraw-Hill, New York, 1973, p.126.

14. L. H. Brixner, *J. Am. Chem. Soc.*, **81**, 3841 (1959).

15. F. R. Bichowsky and F. D. Rossini, *The Thermochemistry of the Chemical Substances*, Reinhold, New York, 1936, p. 126.

16. R. Roy, *J. Am. Ceram. Soc.*, **52**, 344 (1969).

17. J. Livage, *J. Solid State Chem.*, **64**, 322 (1986).

18. J. Livage, M. Henry, and C. Sanchez, *Prog. Solid State Chem.*, **18**, 259 (1988).

19. S. Klosky and C. Marzano, *J. Phys. Chem.*, **29**, 1125 (1925).

20. S. Klosky, *J. Phys. Chem.*, **34**, 2621 (1930).

21. C. B. Hurd, W. J. Jacober, and D. W. Godfrey, *J. Am. Chem. Soc.*, **63**, 723 (1941).

22. W. O. Milligan and H. B. Weiser, *J. Phys. Chem.*, **40**, 1095 (1936).

23. H. B. Weiser and W. O. Milligan, *J. Phys. Chem.*, **40**, 1 (1936).

24. L. L. Hench and D. P. Ulrich (eds.) *Ultrastructure Processing of Ceramics, Glasses and Composites*, John Wiley, New York, 1984.

25. B. E. Yoldas, *J. Mater. Sci.*, **21**, 1087 (1986).

26. J. L. Woodhead, *J. Phys. Colloq.*, C1-47 (1986).

27. R. K. Iler, *The Chemistry of Silica*, John Wiley, New York, 1979.

28. J. P. Jolivet, R. Massart, and J-M. Fruchart, *Nouv. J. Chim.* 7(5), 325 (1983).

29. R. Massart, *IEEE Trans. Magn.*, **MAG-17**(2), 1247 (1981).

30. J. Rouxel, personal communication.

31. J. Rouxel, *Chem. Scripta*, **28**, 33 (1988).

32. J. Livage, *Chem. Scripta*, **28**, 9 (1988).

33. *Rev. Chim. Min.*, **21**(Nos. 1–6) (1984).

34. R. Marchand, L. Brohan, and M. Tournoux, *Mat. Res. Bull.*, **15**, 1129 (1980).

35. F. Théobald, R. Cabala, and J. Bernard, *J. Solid State. Chem.*, **17**, 431 (1976).

36. C. Delmas, Y. Borthomieu, and C. Faure, *Solid State Ionics*, **32/33**, 104 (1989).

37. M. Potel, P. Gougeon, R. Chevrel, and M. Sergent, *Rev. Chim. Min.*, **21**, 509 (1984).

38. M. A. Verneuil, *C.R. Acad. Sci. Paris*, **135**, 791 (1902).

39. S. Zerfoss, L. R. Johnson, and P. H. Egli, *Disc. Faraday Soc.*, **5**, 168 (1949).

40. D. C. Stockbarger, *J. Opt. Soc. Am.*, **39**, 731 (1949).

41. H. E. Buckley, *Crystal Growth*, John Wiley, New York, 1951.

42. H. R. Harrison, R. Aragón, J. E. Keem, and J. M. Honig, *Inorganic Syntheses*, Vol. XXII. John Wiley, New York, 1983, pp. 43–48.

43. P. J. Remeika, *J. Am. Chem. Soc.*, **76**, 940 (1954).

44. P. J. Remeika, *J. Am. Chem. Soc.*, **78**, 4259 (1956).

45. Z. Hauptman, *Czech. J. Phys.*, **12B**, 148 (1962).

46. L. S. Darken and R. W. Gurry, *J. Am. Chem. Soc.*, **68**, 798 (1946).

47. R. Kershaw and A. Wold, *Inorganic Syntheses*, Vol. X. McGraw Hill, New York, 1968, pp. 10–14.

48. M. L. Gray, R. Kershaw, W. Croft, K. Dwight, and A. Wold, *J. Solid State Chem.*, **62**, 57 (1986).

49. F. J. Morin, *Phys. Rev. Lett.*, **3**(1), 34 (1959).

50. M. Pouchard and J-C. Launay, *Mat. Res. Bull.*, **8**, 95 (1973).

51. H. Schäfer, *Chemical Transport Reactions*, Academic Press, New York, 1964.

52. A. A. Ballman and R. A. Laudise, in *The Art and Science of Growing Crystals*, J. J. Gilman, ed. John Wiley, New York, 1963, p. 231.

53. H. Rau and A. Rabenau, *J. Cryst. Growth*, **3/4**, 417 (1968).

54. A. Rabenau, *Angew. Chem. Int. Ed. Engl.*, **24**, 1026 (1983).

55. H. Y. Chen, D. M. Miller, J. E. Hudson, and C. J. A. Westenbroek, *I.E.E.E. Trans. Magn.*, **20**, 24 (1984).

56. O. Glemser, *Angew. Chem.*, **73**, 785 (1961).

57. A. H. Heuer, and L. W. Hobbs, *Advances in Ceramics*, Vol. 3, American Ceramic Society, Columbus, OH. 1981, p. 455.

58. M. L. Bayard, T. G. Reynolds, M. Vlasse, H. L. McKinzie, R. J. Arnott, and, A. Wold, *J. Solid State Chem.*, **3**, 484 (1971).

59. A. W. Sleight, *Inorg. Chem.*, **8**, 1764 (1969).

60. J. W. Pierce, H. L. M. McKinzie, M. Vlasse, and A. Wold, *J. Solid State Chem.*, **1**, 332 (1970).

61. L. R. Rothrock in *Kirk-Othmer Encyclopedia of Chemical Technology*, Vol. 4, 3rd ed. John Wiley, New York, 1980, p.719.

62. R. M. Barrer, *Hydrothermal Chemistry of Zeolites*, Academic Press, New York, 1982.

63. W. Paul and D. M. Warschauer, *Solids under Pressure*, McGraw-Hill, New York, 1963.

64. M. L. Andreiux, *Ann. de Chim.* **12**(10), 423 (1929).

65. M. L. Andreiux, *C. R. Acad. Sci. Paris*, **189**, 1279,(1929).

66. J. L. Andrieux and H. Bozon, *C. R. Acad. Sci, Paris*, **228**, 565 (1949).

67. J. L. Andreiux and H. Bozon, *C. R. Acad. Sci. Paris*, **230**, 952 (1950).

68. E. F. Bertaut and P. Blum, *J. Phys. Radium*, **17**, 175 (1956).

69. E. F. Bertaut and P. Blum, *Acta Crystallogr.*, **9**, 121 (1956).

70. N. F. H. Bright, J. F. Rowland, and J. G. Wurm, *Can. J. Chem.*, **36**, 492 (1958).

71. M. Dodero and C. Déportes, *C. R. Acad. Sci.*, **242**, 2939 (1956).

72. E. Banks and A. Wold, in *Preparative Inorganic Reactions*, Vol. 4, W. L. Jolly, ed. Interscience, New York, 1968, p. 237.

73. P. Hagenmuller, in *Progress in Solid State Chemistry*, Vol. 5, H. Reiss, ed. Pergamon, New York, 1971, p. 71.

74. A. Manthiram and J. Gopalakrishnan, *J. Less-Common Met.*, **99**(1) 107 (1984).

Additional References

D. Elwell and H. J. Scheel, *Crystal Growth from High Temperature Solutions*, Academic Press, New York, 1975.

L. L. Hench and J. K. West, *Chem. Rev.*, **90**, 33 (1990).

Problems

P6.1. Indicate the technique(s) that can be used to prepare the following substances:
   a. VO_2 single crystal
   b. Fe_3O_4 single crystal
   c. Ruby single crystal
   d. Polycrystalline CrN
   e. Diamond thin film
   f. Anhydrous nickel (II) chloride
   g. Anatase
   h. TiB_2
   i. Quartz crystal
   j. ZrO_2 single crystal

P6.2. Choose a single solid state material that has found widespread use as (or is incorporated in) an electronic or optical device.
   a. How is the material prepared in the form necessary for the specific application?
   b. What methods of characterization were used to enable optimization of the desirable characteristic for the application chosen?
   c. What aspect of the crystal chemistry of the material played an essential role in maximizing the desired properties?

P6.3. For each of the following structure types
   a. perovskite
   b. thiospinel
   c. chalcopyrite
   d. ilmenite
   e. layered dichalcogenide
   f. 1-2-3 superconductor

   name a compound that involves one of the following species: Fe(III), Co(III), Cr(III), Ti(IV), Mo(IV), Cu(III) and discuss:
   a. preparation of homogeneous, uniform polycrystalline samples
   b. preparation of well-formed single crystals
   c. chemical analysis and X-ray analysis
   d. magnetic properties
   e. electrical properties
   f. why this compound was studied, and the significant results of such studies.

7

Binary Oxides

A. Transition Metal Monoxides

a. Transition Metal Monoxides with the Rock Salt Structure

The $3d$ transition metal monoxides, TiO, VO, ___, MnO, FeO, CoO, NiO, and CuO are a most difficult group of compounds to prepare stoichiometrically and characterize. Whereas TiO and VO are metallic, MnO, FeO, CoO, and NiO are semiconductors. There is no reliable evidence for the existence of CrO. The transition metal monoxides represent an interesting class of compounds that present many of the classic problems faced by solid state chemists and hence represent an excellent starting point for the treatment of the preparation, structure, and properties of transition metal oxides.

The oxide chemistry of Ti and V is quite complex and Wells (1) summarizes clearly many of the structural problems associated with these two metal oxide systems. In this treatment, the choice of oxides in both the Ti-O and V-O systems will be limited to specific phases in which both the chemistry can be controlled, and where the compounds selected show electronic properties which can be related to their structure.

TiO can best be prepared by the reduction of TiO_2 with metallic titanium. Watanabe et al. (2) prepared TiO samples by heating mixtures of Ti (obtained by decomposing TiI_4) and high purity TiO_2 in a tungsten arc furnace under an argon atmosphere for 50 hours. Oxygen concentrations were determined by converting the products to TiO_2 in oxygen at 1200°C and determining the weight changes by thermogravimetric analysis. The accuracy of these analyses was estimated at 0.5% oxygen. The low temperature phase of TiO was found to be stable in the composition range $TiO_{0.9}$–$TiO_{1.1}$ below 990°C. Terauchi and Cohen (3), grew crystals of TiO in a purified argon atmosphere in a tri-arc furnace. The structure of TiO is related to NaCl but has an ordered array of vacant lattice sites. The unit cell is monoclinic (space group $B2/b$) with $a = 9.340(5)$ Å, $b = 5.860(4)$ Å, $c = 4.141(1)$ Å and $\gamma = 107.553°$. There are Ti and O ion vacancies every third (100) plane in the [010] direction of the monoclinic structure.

In this oxide, the cation–cation interactions are stronger than any cation–anion–cation interactions. It was proposed by Morin (4) and Goodenough (5) that the conduction band is produced by the overlap of partially filled d-orbitals of neighboring Ti(II) cations. These d-orbitals are split by the cubic crystalline field present at the cation sites into lower lying t_{2g} and higher energy e_g levels. The partially filled t_{2g} orbitals of neighboring Ti(II) ions overlap to form a partially filled band that gives rise to metallic behavior. As has been indicated, TiO has a considerable concentration of both cation and anion defects. These defects tend to further reduce the strength of the antibonding metal–oxygen–metal π^* interactions relative to the direct metal–metal bonding interactions. TiO has been reported to become superconducting at approximately 2 K (6).

Much less research has been reported on the properties of VO. Morin (4) has shown that there is a transformation in the measured electrical resistivity at $\approx 150°C$ (Fig. 7.1). These studies were carried out on single crystal samples 0.1 mm in size. These were too small for four-probe measurements, so two-probe contacts were used. The crystals studied were grown hydrothermally at the Bell Laboratories but the specific procedure was not published. The observed transition was sharp with no measurable time or temperature dependence. VO was observed to be an insulator below the transition temperature and the conductivity was observed to increase by a factor of 10^6 above the transition. VO exhibits both anionic and cationic vacancies and Geld et al. (7–10) showed by X-ray analysis, pycnometry, and microscopy that the compositions $VO_{0.75}$ and $VO_{1.30}$ consist of two phases. This is essentially in agreement with the work reported by Andersson (11) and Westman and Nordmarx (12). It was also claimed (7–10) that the composition $VO_{1.30}$ was actually an equilibrium of VO and V_2O_3.

Both Morin (4) and Goodenough (5) have indicated that strong cation-cation interactions are responsible for the observed metallic-type conductivity above the transition point. Below the transition, VO is a semiconductor. From Fig. 7.1 it is readily seen that the electrical properties of VO resemble those of V_2O_3 rather than TiO. In fact, the magnitude of the transition of VO is about the same as that of V_2O_3. From the studies of Geld et al., it is conceivable that the reported transition was measured on a crystal whose composition was not VO, and this compound should be reexamined.

Stoichiometric dark green MnO can be formed by the decomposition of $MnC_2O_4 \cdot 2H_2O$ under a stream of nitrogen at 450°C followed by heating in an oxygen free hydrogen atmosphere at 800°C. Manganese(II) carbonate can also be used as the starting material. The stability of the green MnO phase to oxygen depends on the temperature of preparation. Low-temperature decomposition of the carbonate or oxalate gives a reactive phase that oxidizes readily to $MnO_{1.13}$ (1). The stoichiometric compound crystallizes with $a = 4.442(1)$ Å. At 80 K,

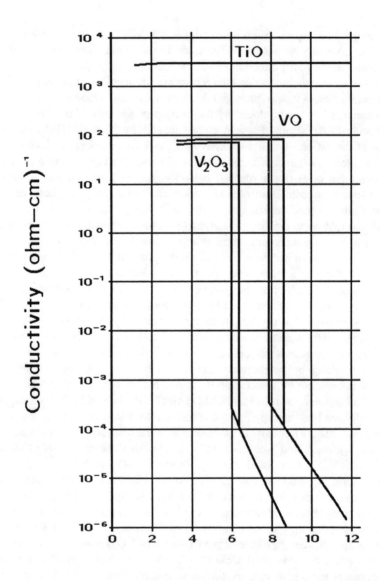

Figure 7.1. Conductivity as a function of reciprocal temperature for the lower oxides of titanium and vanadium.

MnO shows an antiferromagnetic ordering and from neutron diffraction data (13) $a = 8.85$ Å for the magnetic cell. This doubled magnetic cell is a result of the antiferromagnetic ordering of the manganese $3d^5$ ions. This type of ordered cell is not to be confused with superstructures resulting from the ordering of two or more kinds of atoms rather than their random arrangement.

The binary oxide FeO apparently does not exist at normal pressures (14). It has been reported (15–17) that the deviation from stoichiometry, i.e., $Fe_{1-x}O$ results from vacancies on cation sites, and some of the tetrahedral sites are occupied by cations. A careful structure determination was carried out by Koch and Cohen (18) on a single crystal of $Fe_{0.902}O$. The starting material for this crystal was zone refined iron (99.986) equilibrated in a CO/CO_2 (2:3) atmosphere at 1000°C. The analysis was established by the gain in weight on oxidation. From this structural analysis, it was concluded that there were periodically spaced clusters of vacancies. Each cluster of vacant neighboring octahedral cation sites was grouped about occupied tetrahedral cation sites. These clusters did not appear to be regions of magnetite; rather each cluster appeared to consist of 13 vacancies and 4 tetrahedral ions.

Cobalt(II) oxide is an olive-green powder that is readily oxidized by air at room temperature. On oxidation, the compound becomes brown and then black. To minimize the air oxidation of CoO, the compound is prepared by heating cobalt carbonate in a platinum boat under a purified nitrogen atmosphere at 1000°C. The high-temperature preparation has a metal to oxygen ratio approaching 1:1 and is reasonably stable toward air oxidation.

Cobalt(II) oxide crystallizes with the sodium chloride structure $a = 4.26$ Å (19). Redman and Steward (20) reported the preparation of both a hexagonal wurtzite form of CoO (with $a = 3.2$ Å and $c = 5.2$ Å) and a cubic zincblende form ($a = 4.5$ Å). However, these phases appear to be stabilized by impurities present in the compounds. CoO is antiferromagnetic, with a Néel point of 292 K (21). The magnetic transformation is accompanied by a tetragonal distortion of the sodium chloride structure (22–25), which increases with decreasing temperatures.

As one proceeds from TiO to NiO, the $3d$ band formed by the overlap of transition metal t_{2g} orbitals becomes extremely narrow so that by MnO the $3d$ charge carriers occupy energy levels localized on the cations. Pure stoichiometric NiO is a semiconductor with a room temperature resistivity greater than 10^{10} Ω−cm. However, small deviations from stoichiometry or the substitution of monovalent lithium for nickel results in a marked lowering of the resistivity. For conduction in the $3d$ levels to occur, there must be the introduction of Ni(III) species resulting from cation vacancies (or chemical substitution). There can then be an electron exchange between Ni(III) and Ni(II) ions in the rock salt structure. Hence, nickel oxide can be made conducting at room temperature by introducing Ni(III) into the lattice either by a departure from stoichiometry or by the substitution of lithium for nickel(II) ions.

b. Copper(II) Oxide

Copper(II) oxide occurs in nature as the mineral tenorite. The structure (26) shows Cu(II) with square planar coordination of oxygen around the copper. The space group is $C2/c$ with unit cell dimensions $a = 4.6837(5)$ Å, $b = 3.4226(5)$ Å, $c = 5.1288(6)$ Å, and $\beta = 99.54(1)°$. Figure 7.2 shows the chains of oxygen coordination parallelograms formed by sharing edges. The structure is a distorted PdO type with two O-Cu-O angles of 84.5° and two of 95.5°. Each copper has four O' neighbors at 1.96 Å and two next nearest O" neighbors at 2.78 Å. The line O"-Cu-O" is inclined at 17° to the normal to the Cu(O')$_4$ plane and the shortest Cu-Cu distance is 2.90 Å.

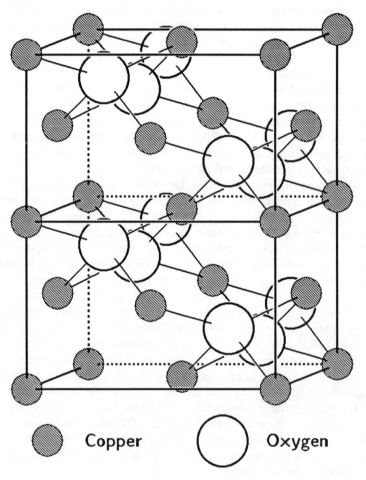

Copper   Oxygen

Figure 7.2. Structure of CuO.

Another oxide of copper, $Cu_{16}O_{14}$ has been reported (27). It is the rare mineral paramelaconite and crystallizes with the space group $I4_1/amd$, with unit cell dimensions $a = 5.817$ Å, $c = 9.803$ Å. The structure is related to tenorite but the oxygen vacancies result in a rearrangement of the monoclinic tenorite structure to a tetragonal one.

In Cu(II) oxide, the Jahn-Teller distortion due to the stabilization of a single d-hole per atom, $Cu(II)3d^9$, results in the observed difference of its structure with the monoxides of the other first row transition metals. For CuO, the equitorial oxygen p_σ overlaps strongly with the $d_{(x^2-y^2)}$ orbital to form a low-lying σ band and a high-energy σ^* band, but little, if any, with the $d_{(z^2)}$, which remains non-bonding. The copper d_{xy}, d_{yz}, d_{zx} orbitals contribute to the formation of bands which overlap the σ^* bands. O'Keefe and Stone (28), Roden et al. (29), as well as Yang et al. (30) indicated that CuO is antiferromagnetic with an observed anomaly in the magnetic susceptibility at 230 K. In stoichiomatic CuO, the σ^* band of Cu(II) is half filled. The Goodenough band model (31,32) indicates that when antiferromagnetic behavior is observed in the copper oxides, a separation may occur between empty σ^* band states and occupied σ^* band states. Goodenough designated these empty and occupied states as the Cu(II/I) couple and the Cu(III/II) couple, respectively. This is shown in Fig. 7.3 where the Fermi energy is placed near the top of the filled π^* band. In this figure, the relative density (per unit energy) of electronic states in each band is plotted as a function of electron energy. The antiferromagnetic behavior of CuO is not typical, since

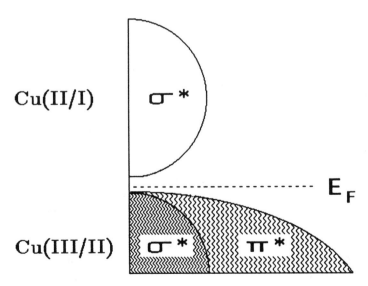

Figure 7.3. Schematic band structure for CuO, showing the relative density (per unit energy) of electronic states in each band as a function of electron energy.

the data reported by Roden et al. (29) do not show a maximum of χ at the antiferromagnetic ordering temperature. The unusual ordering can be related to the magnitude, or even existence, of the σ^* bandsplitting. In the case where Cu(III) is formed, the sign of the carrier results unambiguously from the formation of holes in the top of the π^* band. Where Cu(I) is formed, in addition to any electrons entering the σ^* Cu(II/I) band, holes in the π^* band are also created by thermal excitation. Since the σ^* bands are narrow, the greater mobility of the holes in the π^* band will determine the sign of the dominant carrier. Koffeyberg and Benko (33) have indicated that CuO can only be prepared as p-type.

B. Transition Metal Dioxides

The rutile structure was first determined by Végard (34). The structure is shown in Fig. 7.4 and the details of the structure will be discussed shortly using TiO_2 as an example. In nature, a number of minerals crystallize with the rutile structure, namely, MnO_2, TiO_2, SnO_2, and PbO_2. The atomic arrangement found in rutile in which the cation is in six coordination is found in a large number of fluorides and oxides, e.g., MgF_2, ZnF_2, MnF_2, FeF_2, NiF_2, VO_2, NbO_2, MoO_2, WO_2, RuO_2, OsO_2, IrO_2. Whereas FeOF, TiOF, VOF, and MgH_2 also crystallize with this structure, the other dihalides, disulfides do not. Difluorides and dioxides, which contain larger ions, are more apt to adopt the fluorite rather than the rutile structure.

a. Titanium(IV) Oxide

The oxide TiO_2 exists in three crystalline modifications at atmospheric pressure, namely rutile, anatase, and brookite. All have been prepared synthetically and characterized. Pure polycrystalline TiO_2 prepared at high temperature crystallizes with the rutile structure (space group $P4_2/mnm$). The product may be obtained by hydrolysis of pure $TiCl_4$ and final heating to 1000°C (35,36). The structure of TiO_2 rutile is shown in Fig. 7.4. The metal coordinates are Ti:(000),(1/2,1/2, 1/2) and O:$\pm(xx0)$, $\pm(1/2 + x,1/2 - x,1/2)$. As can be seen from Fig. 7.4, the structure consists of chains of TiO_6 octahedra and each pair share opposite edges. Each titanium atom is surrounded octahedrally by six oxygen atoms, whereas each oxygen is surrounded by three titanium atoms arranged as corners of an equilateral triangle. The structure, therefore, has a 6:3 coordination. A number of transition metal dioxides crystallize with rutile-related structures. Magnéli et al. (37–40), as well as Goodenough (41,42), have related certain properties of various transition metal dioxides to the rutile structure and to the number of transition metal d-electrons.

Goodenough proposed a one-electron energy diagram for rutile which is shown in Fig. 7.5. Rogers et al. (43) applied the Goodenough model to construct a diagram by mixing $e_g^2 sp^3$ wave functions of titanium from a doubled formula

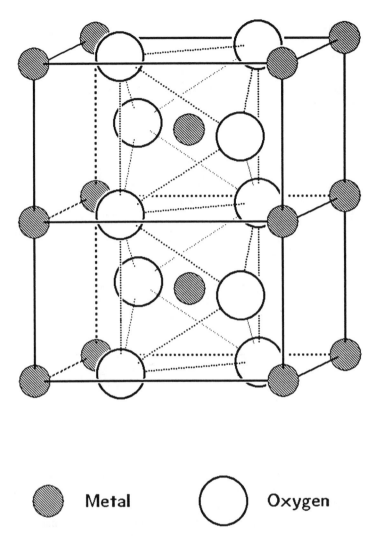

Figure 7.4. The structure of rutile.

unit, Ti_2O_4, with hybridized sp^2 wave functions of oxygen to form bonding σ and antibonding σ^* states. As a result of the long-range ordering of the structure, the states are broadened to form bands with finite widths. Interaction between two t_{2g} cation orbitals ($t_{2g\perp}$) with the remaining oxygen p-orbital gives rise to bonding and antibonding π bands. The $t_{2g\parallel}$ orbitals of the cations are directed along the c axis and may form a σ band if the metal-metal separation is small enough for significant overlap to occur. The allowable number of electrons that can be accommodated by any band is given in brackets and this number is equal to the

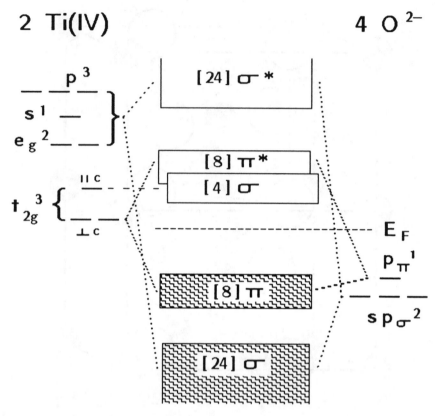

Figure 7.5. Schematic energy bands for TiO_2.

product of the number of atoms per formula unit, the orbital degeneracy per atom, and the spin degeneracy per orbital. The rutile TiO_2 has an electronic configuration of $3d^0$ for Ti(IV). The Ti–O σ and π bands are completely filled so that the Fermi level is located at the middle of the energy gap between those bands and the antibonding π^* and cation–cation d-bands. Hence, TiO_2 should be a semiconductor and become metallic when the Fermi level is raised as the TiO_2 is reduced to form some Ti(III) $3d^1$ states.

This model was useful in understanding the electronic properties of reduced TiO_2 samples. In a study reported by Subbarao et al. (44), 1-mm wafers of a TiO_2 crystal were heated in a stream of pure hydrogen between 500° and 700°C. The oxygen content in the resulting samples was determined by thermogravimetric analysis. Ground samples were heated in an oxygen atmosphere at 1000°C and the weight gain was measured. The variation of oxygen content, i.e., x in TiO_{2-x} with reduction temperature, is given in Fig. 7.6. Resistivity measurements were obtained on the reduced wafers by the standard van der Pauw technique. Care

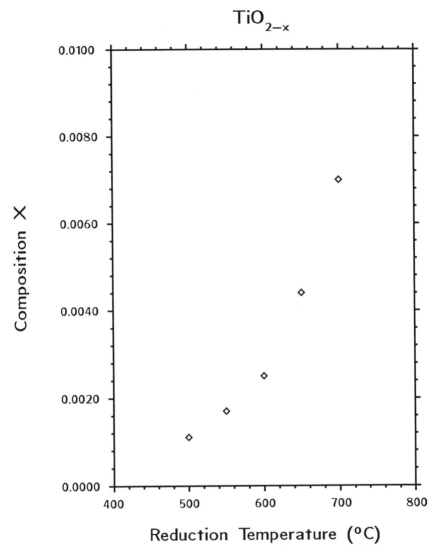

Fig. 7.6. Variation of composition x with the reduction temperature of TiO_{2-x}.

was taken to verify uniform conductivity throughout the wafer. The measured resistivity values are shown as a function of reduction temperatures in Fig. 7.7.

The rutile TiO_2 is an insulator, the Ti(IV) ion having a $3d^0$ electronic configuration. However, with small additions of Ti(III) ($3d^1$) electrical conductivity can simultaneously occur via both direct Ti(III)–Ti(IV) and Ti(III)–O–Ti(IV) interactions (43). Whereas conductivity can be produced in the rutile TiO_2 by the introduction of Ti(III), i.e., the formation of TiO_{2-x}, there has been evidence

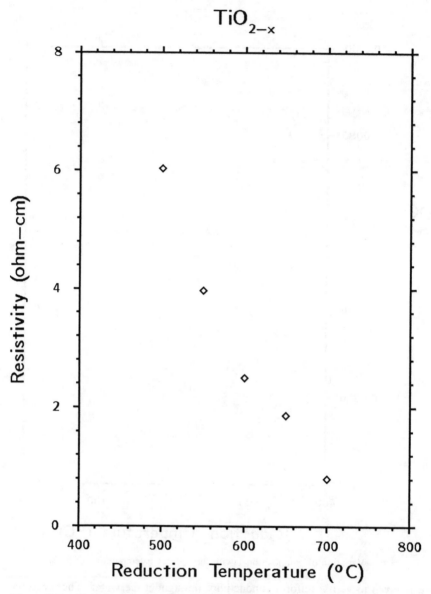

Figure 7.7. Variation of resistivity with the reduction temperature of TiO_{2-x}.

(45,46) that these compounds are not stable in the presence of oxygen at their surfaces. An alternative method of producing conducting compounds is to substitute fluorine for oxygen rather than to create oxygen vacancies. Both methods result in the formation of $3d^1$ titanium, which would account for the relatively high conductivity obtained. Fluorination of TiO_2 was reported to occur with HF produced by the decomposition of KHF_2 at 260°C (47). The HF was carried over the sample with an argon/hydrogen (85/15) mixture at 575–700°C. The fluoride content of the sample at 700°C gave a value of $x = 0.002$ in $TiO_{2-x}F_x$, whereas that of the sample fluorinated at 600°C gave a value of $x = 0.0001$, which was the limit of fluoride detection. For the sample where $x = 0.002$, the measured resistivity was 2.5 Ω–cm.

b. Vanadium (IV) Oxide

V_2O_3 crystallizes with the corundum structure whereas VO_2 is a rutile. Between these two oxides, Andersson (11) and Andersson (48) found a family of compounds with the stoichiometry of $V_nO_{2n-1}(3<n<8)$. The term "shear structures" was introduced by Wadsley (49) to describe these intermediate oxides that are formed from slabs of rutile-like structure sharing faces. Most of these phases are triclinic except V_3O_5, which is monoclinic.

The vanadium oxides with compositions between V_2O_3 and VO_2 are formed from slabs of a rutile-like structure sharing faces. Each of these phases shows transitions from the semiconducting to metallic state. Many of their properties can also be explained by considering structural changes that occur at the transition and by utilizing the concepts of Goodenough (5) in which strong covalent mixing with oxygen p_π orbitals to form a π^* band plays a central role (see Fig. 7.8). For

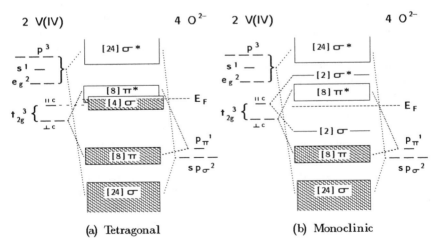

Figure 7.8. Schematic energy bands for (a) tetragonal and (b) monoclinic VO_2.

VO_2, above the transition temperature of 67°C, these bands overlap and are partially filled, giving rise to metallic conductivity. The compound is tetragonal ($P4_2/mnm$) and all the V$-$V distances are equal (2.87 Å). Below the transition temperature, the formation of V$-$V pairs along the tetragonal c axis causes a doubling of the crystallographic unit cell (50) and a splitting of the $t_\parallel$ band in two. The resulting lower $t_\parallel$ band lies below the bottom of the π^* band, becomes filled, and lowers the Fermi energy below the bottom of the remaining π^* bands. The π^* band, arising from covalent mixing, is now empty (51). Excitation of electrons from the filled $t_\parallel$ band to the bottom of the π^* gives rise to the semiconducting behavior of VO_2 below 67°C (Fig. 7.8).

Bayard et al. (51) studied the effect of fluorine substitution for oxygen in VO_2. Single crystals were synthesized using hydrothermal techniques described by Pierce et al. (52). The procedure involves the reaction of V_2O_5 and V in 6 percent aqueous HF. The synthesis is carried out in sealed gold tubes at 600°C and 1.33 kbar pressure. The oxyfluoride crystals that were obtained had the composition $VO_{2-x}F_x(0<x<0.2)$. For VO_2, the monoclinic to tetragonal transition temperature decreases with increasing anion substitution, i.e., the high temperature tetragonal phase becomes more stable (53).

For the system $VO_{2-x}F_x$, activation energies were calculated from the resistivity data (see Fig. 7.9), and the values are given in Table 7.1 for different values of x in $VO_{2-x}F_x$. It can be seen from these curves that a metallic to semiconductor transition occurs at a temperature that decreases with increasing values of x. This change in the electrical properties of $VO_{2-x}F_x$ can be explained by the corresponding transition from the monoclinic phase to the tetragonal phase, which has been observed by means of low-temperature X-ray analysis. A linear relationship exists between the value of x and the transition temperature (T_t) as shown in Fig. 7.10; this extrapolates to the correct transition temperature for pure VO_2. For compounds with higher fluorine content, the transition region is broadened considerably, and the transition points T_t were chosen as the first deviation from log-linear behavior. A similar linear relationship exists between the volume of the tetragonal cell and the value of x, again extrapolating to the value of the pure VO_2 phase at $x = 0$ (see Fig. 7.11). The same behavior has been observed in compounds corresponding to the formula $V_{1-x}W_xO_2(0 \leq x \leq 0.67)$ by Nygren and Israelsson (53).

It is seen, therefore, that the addition of fluorine tends to stabilize the high-temperature, higher symmetry, rutile phase. The compositions containing large amounts of substituted fluorine show primarily metallic behavior. However, for all compositions studied, there still appears to be a discontinuity in the resistivity at the expected transition temperature (T_t). The increased substitution of fluorine for oxygen in the system $VO_{2-x}F_x$ results in the creation of additional unpaired d-electrons. Despite the tendency for the more electronegative fluoride anion to localize d-electrons, it is apparent that for the amount of fluorine that has been substituted, the conduction paths have not been substantially blocked.

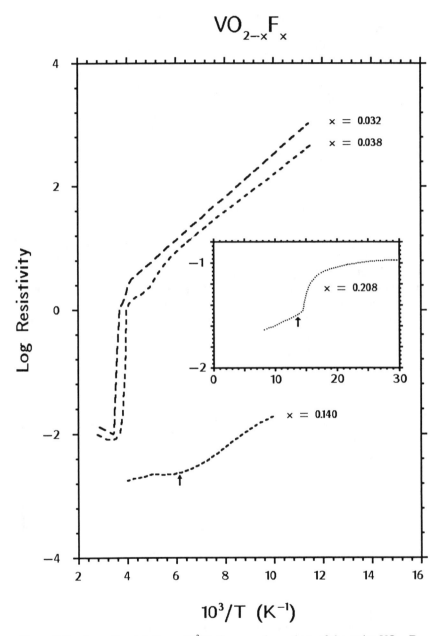

Figure 7.9. Log of resistivity vs $10^3/T_K$ for several members of the series $VO_{2-x}F_x$.

Table 7.1. Cell Parameters of Members of the System $VO_{2-x}F_x$

Compound	Cell parameters (Å)		Activation energy (eV)	$T_t(K)$
	a	c		
VO_2	4.530 ± 4	2.869 ± 3	0.5 just below T_t	340
$VO_{1.97}F_{0.03}$	4.552 ± 4	2.853 ± 3	0.07	298
$VO_{1.96}F_{0.04}$	4.554 ± 4	2.854 ± 3	0.06	282
$VO_{1.86}F_{0.14}$	4.562 ± 4	2.876 ± 5	< 0.01	155
$VO_{1.79}F_{0.21}$	4.569 ± 4	2.886 ± 3	< 0.01	65

c. Other Transition Metal Dioxides

Many of the transition metal dioxides crystallize with the rutile or rutile-related structure. Rogers et al. (43) have shown that the electronic properties and changes in structure can be related to the number of d-electrons. The magnetic properties vary from Pauli paramagnetic to ferromagnetic, and the electrical resistivities range from semiconducting to metallic.

The dioxides of Ti, Mn, Ru, Os, Ir, and Pt can be prepared by direct oxidation of the metals, lower oxides, chlorides, or by decomposition of the nitrates (35,43,54,55). The synthesis of RhO_2 and PtO_2 can be achieved only at high pressures (55). The dioxides, VO_2, NbO_2, MoO_2, WO_2, and β-ReO_2, can be synthesized by reduction of higher oxides (35,56–58). Chromium dioxide is prepared by the hydrothermal reduction of CrO_3 (59).

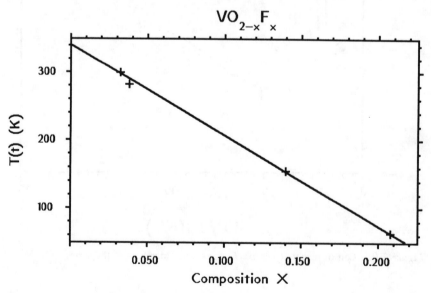

Figure 7.10. T_t vs composition x for $VO_{2-x}F_x$.

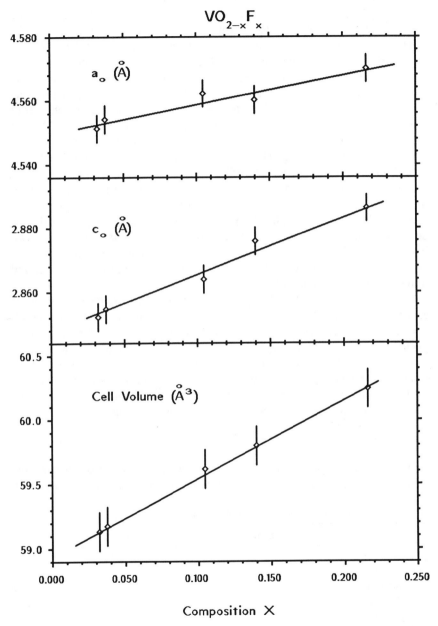

Fig. 7.11. Cell constants a_o, c_o *and V vs composition x in* $VO_{2-x}F_x$.

MoO_2 and WO_2 crystallize with the monoclinic structure of the low temperature form of VO_2. Wells (60) tabulated the metal-metal distances along the c axis for a number of transition metal dioxides and they are given in Table 7.2. The monoclinic distortion results in long and short distances. The low-temperature form of NbO_2 has a tetragonal superstructure of the rutile type. In the rutile-like structures, close approach of metal atoms or strong metal-oxygen-metal overlap can give rise to metallic behavior. However, structure, as well as the number of d-electrons, can affect markedly the electronic properties of these oxides. Magnéli and co-workers (37–40,61) indicated that for MoO_2 and related oxides, there is strong metal–metal bonding. The axial ratios determined for these oxides can be related to the number of free electrons per bond. A simple model has been proposed by Goodenough (41,42) and applied by Rogers et al. (43) to account for the electronic behavior of these oxides. MoO_2 has 2 unpaired electrons for each molybdenum atom. Therefore, one electron is available for Mo-Mo σ bonding and the second electron can partially fill the Mo–O π^* band (Fig. 7.12). However, the short Mo–Mo separation is not fully consistent with a single σ bond. Hence, some of the electrons from the metal–oxygen band are transferred to metal–metal π bonding (43).

For CrO_2, Rogers et al. (43) indicated that there is a sharp decrease in direct cation–cation bonding. Hence, the $t_{2g \parallel c}$ do not enter into metal-metal σ bonds but remain at discrete cation sites in this compound (Fig. 7.13). A number of the platinum metal dioxides, e.g., Ru, Ir, Os, and possibly Rh, show metallic behavior that is consistent with partial filling of the M–O π^* band. For PtO_2, where the π^* band is filled, the Fermi level is raised to be between the π^* and σ^* bands, and the compound should behave as a semiconductor. Rogers et al. (43) reported powder samples of PtO_2 to be semiconducting.

d. The Crystal Chemistry of ZrO_2

ZrO_2 crystallizes with the fluorite structure and this oxide is of considerable technical interest. It has a high melting point, low thermal conductivity, high

Table 7.2. Metal-Metal Distances in Some Dioxides

Oxide	No. of electrons	M–M separation in chain (Å)	
TiO_2	0	2.96	
VO_2 (tetragonal)	1	2.85	
VO_2 (monoclinic)		2.62	3.17
NbO_2 (tetragonal)	1	(3.0)	
(superstructure)		2.71	3.30
MoO_2	2	2.51	3.10
WO_2	2	2.48	3.10
RuO_2		3.11	
OsO_2		3.18	

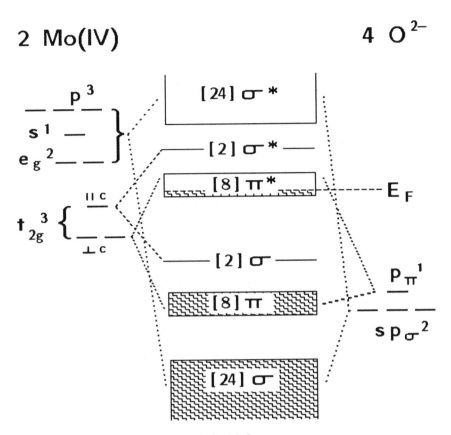

2 Mo(IV)   4 O²⁻

Figure 7.12. Schematic energy bands for MoO₂.

corrosion resistance, and good mechanical properties; hence, ZrO₂ can be used both as a refractory and ceramic material. Because of its remarkable strength and toughness, it has been called "ceramic steel" (31). ZrO₂ also shows ionic conductivity through oxygen migration, and can therefore be employed as ionic conductors and oxygen sensors to measure or monitor oxygen partial pressures. In addition, since stabilized zirconia has a refractive index $n = 2.15$, the crystals are used as diamond substitutes (32). Zirconia is also a good catalyst support because of its ability to interact with various transition metal oxides (63–66). Alexander and Bugosh (67) hydrolyzed acidic solutions of zirconium oxychloride at 120° and 150°C for 1 hour and obtained zirconium oxide particles 50–100 Å in size with 5–400 m²/g surface area. These particles are hydrated and can be represented as ZrO₂·nH₂O where the water content is variable.

Polymorphism of ZrO₂ is now well known. It exhibits three crystalline phases related to the fluorite structure. The monoclinic phase is the thermodynamically stable phase up to 1100°C. The tetragonal phase exists in the temperature range

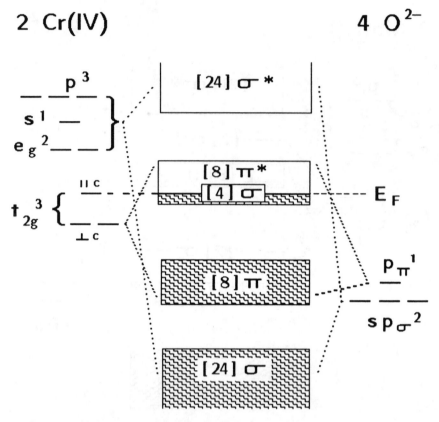

Figure 7.13. Schematic energy bands for CrO_2.

of 1100–2370°C, and the cubic phase is found above 2370°C (68). The transition at 1100°C restricts the use of ZrO_2 as a refractory material. Continued thermal cycling through this transition temperature causes cracking and disintegration. The shattering of ZrO_2 is primarily due to a volume change of ≈9% that accompanies the monoclinic-to-tetragonal transition. If solid solutions of CaO, MgO, or Y_2O_3 with ZrO_2 are prepared, the high-temperature cubic polymorph of ZrO_2 is stabilized and the phase transition as well as the shattering is avoided (69,70). The replacement of the Zr^{4+} cation by a cation with a lower charge results in the creation of anion vacancies (or interstitial cations) in order to maintain charge balance. The substitution of Ca^{2+} for Zr^{4+} can be represented as $Ca_xZr_{1-x}O_{2-x}$. Stabilized zirconia shows a reversible expansion–contraction behavior on thermal cycling. There is also the disappearance of any structurally destructive phase transitions. The formation of oxide vacancies results in the use of stabilized zirconia as oxygen selective electrodes since, at high temperatures, these vacancies are mobile.

The cubic phase of ZrO_2 is a fluorite-type structure with each zirconium atom surrounded by eight equidistant oxygen atoms and each oxygen atom tetrahedrally coordinated by four zirconium atoms. Figure 7.14 shows the (100) projection of this structure. The tetragonal phase, determined by Teufer (71), is a distorted fluorite structure which is characterized by an 8-fold coordination of zirconium atoms and two different sets of Zr–O distances, which are 2.445 and 2.065, respectively (72). Monoclinic ZrO_2 has 7-fold coordination of zirconium atoms

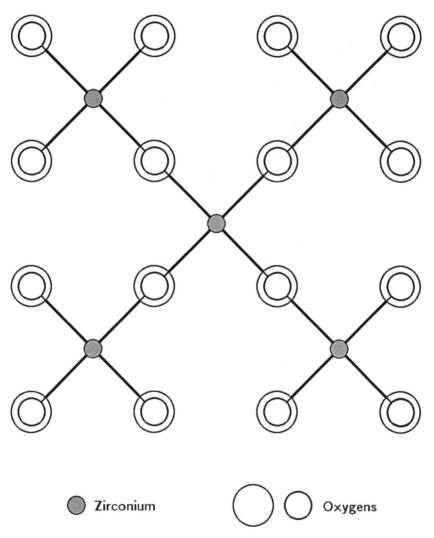

⬤ Zirconium   ◯ ◯ Oxygens

Figure 7.14. One layer of ZrO_8 groups in the ZrO_2 structure. The plane of projection is (100).

and is a layer-type structure consisting of two different sets of oxygen atoms. The O(I) atoms form a planar square group, which would correspond to one-half of a normal 8-fold cubic array. The O(II) atoms form a somewhat irregular triangle whose plane is nearly parallel to the plane of O(I) atoms (Fig. 7.15) (72).

Wolten (73) has shown by single crystal X-ray analysis that the monoclinic structure, space group $P2/c$, can be transformed to a tetragonal structure of space group $P4/nmc$ martensitically through a diffusionless shear mechanism where the atoms retain their neighbors in either phase. The transformation is accompanied

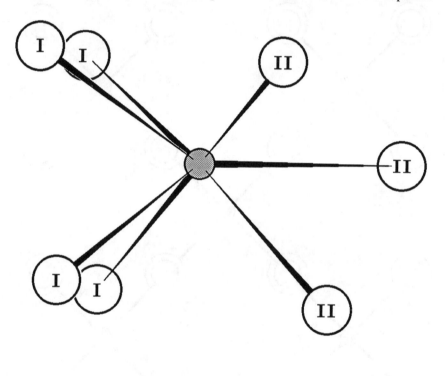

Figure 7.15. The ZrO$_7$ polyhedron in the monoclinic ZrO$_2$ structure.

by a change in coordination of zirconium atoms from seven to eight, which is accomplished by rotation of the O(I) atoms. The martensitic transformation is rapid since it requires neither nucleation nor subsequent growth of a second phase. Therefore, the tetragonal (or cubic) phase cannot be retained by quenching.

Although tetragonal ZrO_2 is stable only above 1100°C, it can be prepared as a metastable phase at much lower temperatures by some special preparative methods. Several investigators have prepared the tetragonal form at low temperatures (74–76). The transformation to the monoclinic phase does not occur appreciably below 600°C. Several explanations have been advanced to account for the low temperature stabilization of tetragonal ZrO_2. Clearfield (77) prepared an amorphous precipitate of hydrous ZrO_2, which was heated in distilled water under reflux. The sequence of phases that he observed in the precipitate was amorphous → tetragonal → tetragonal + monoclinic → monoclinic. Garvie (78) proposed that the stabilization of tetragonal ZrO_2 is correlated with their small crystallite size, large specific surface and appreciable excess energy. Davis (76) demonstrated that the pH of the solution from which the precursor gel is precipitated determined the structure of the final zirconium oxide product. Davis was able to produce monoclinic ZrO_2 from gels precipitated from solutions with a pH range of 6.5–10.4. The tetragonal phase formed from precursors prepared from solutions with pH 3–4 or 13–14. Furthermore, he indicated that the temperature of decomposition of the precursor, the atmosphere used and the presence of alkali did not affect appreciably the zirconium phase compositions.

Zirconium dioxide thin films can be used for optical coatings, protective coatings and insulating layers (79,80). Various fabrication techniques have been employed for the preparation of high-quality films. These include pulsed laser evaporation (81), ion assisted deposition (82), chemical vapor deposition (CVD) (83), and metallorganic chemical vapor deposition (MOCVD) (84). Smooth and homogeneous zirconium oxide thin films were grown on both silicon and silica substrates using an ultrasonic nebulization and pyrolysis technique (85). The films had good adherence to both substrates. These films were transparent and had submicron grain texture with a band gap of 5.68 eV.

C. Several Transition Metal Oxides with the Corundum Structure

A number of binary transition metal oxides crystallize with the corundum $(\alpha-Al_2O_3)$ structure including Ti_2O_3, V_2O_3, Cr_2O_3, and Rh_2O_3. The compounds which will be discussed in this section are Ti_2O_3, V_2O_3, and Rh_2O_3. The first two have interesting electronic properties and Rh_2O_3 is an unusual platinum metal oxide which is rarely discussed in most of the published literature.

The preparation of pure Ti_2O_3 is quite difficult. The reduction of TiO_2 in a stream of hydrogen yields $TiO_{1.6}$ at 1250°C and $TiO_{1.46}$ at 1430°C. An alternative method involves the reaction of finely divided Ti powder (oxygen free) with TiO_2

in sealed evacuated tubes at elevated temperature. However, in this method high temperatures are required and if silica tubes are used, there can be considerable attack of these reaction tubes.

Ti_2O_3 crystallizes with the corundum structure (space group $R\bar{3}c$) and shows a gradual change from semiconductor to metallic behavior in the temperature range from 400–600 K (4,86,87). A number of investigators (88-90) found that at all temperatures, Ti_2O_3 was isomorphous with α–alumina. The c/a ratio at room temperature was reported to be 2.64 (90) and, with heating, this ratio increased to 2.73 at 597°C. Reed et al. (91) prepared the oxide by mixing high-purity titanium powder obtained from the decomposition of titanium hydride with the appropriate amount of TiO_2. The mixture was heated at 1000°C under vacuum for 16 hours and the resulting powder was pressed into pellets and used to grow Ti_2O_3 single crystals. Reed et al. (91) grew large single crystals of TiO_x(1.501 $\leq x \leq$ 1.512) where the total carbon and nitrogen content was below 180 ppm.

The corundum structure, shown in Fig. 7.16, contains TiO_6 octahedra that share one face and three edges with other octahedra. The oxygen ions are coordinated by four cations and are hexagonal close packed; two-thirds of the octahedral sites are occupied (Fig. 7.17). The most widely accepted model for the semiconductor–metal transition for Ti_2O_3 was proposed by Goodenough (92,93) and is shown in Fig. 7.18. For the corundum structure, the a_1 states form two doubly degenerate bands that may be split, and the lower band is filled by a $3d^1$ electron from each of four titanium cations per unit cell.

— A

— B

— A

— B

— A

Figure 7.16. The stacking of octahedrons in the corundum structure.

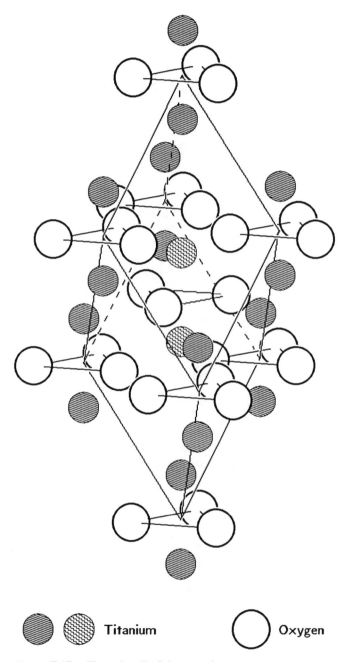

Titanium   Oxygen

Figure 7.17. The unit cell of the corundum structure.

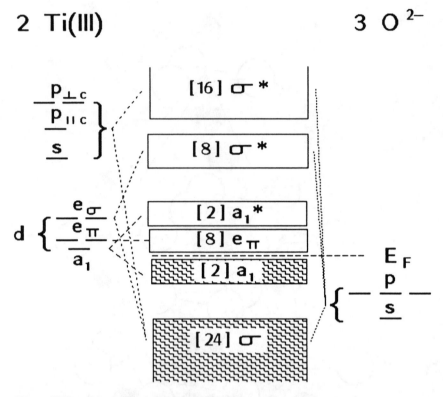

Figure 7.18. Schematic energy bands for Ti_2O_3.

For simplicity, Fig. 7.18 considers only two titanium cations and three oxygen anions as contributing to the formation of bands. Hence, the total electron supply is also halved. It can be seen from Fig. 7.18 that the localized O^{2-} orbitals are more stable than the $3d$ orbitals of the cation. The $3d$ orbitals of the cation are split by crystalline fields into two e_σ orbitals directed toward neighboring anions, two e_π orbitals directed toward the near neighbor cations in the basal plane, and one a_1 orbital directed along the c axis. Covalent mixing of cationic $e_\sigma^2 sp^3$ and anionic sp^3 orbitals results in the formation of bonding σ bands as well as antibonding σ^* bands. Goodenough proposed (93) that the cation–cation interactions are strong enough to transform the e_π orbitals into narrow bands and the a_1 orbitals into bonding and antibonding molecular states localized at discrete c axis cationic pairs. The widths and stabilities of the e_π and a_1 bands that are formed are sensitive to Ti–Ti separations. Covalent mixing with the anion sp^3 orbitals can give rise to 90° cation–anion–cation interactions which transform e_π and a_1 orbitals from cation sublattice orbitals to band-electron orbitals for the entire crystal. The widths and relative stability of the bands depend on the Ti-Ti distances.

The high temperature structural changes and semiconductor–metal transition observed for Ti_2O_3 are consistent with the Goodenough model. At low temperatures the Fermi level falls in the gap between the a_1 and e_π bands (see Fig. 7.18). At higher temperatures, the a_1 band is raised relative to the e_π band, which is broadened as well. The electrons can then be more readily excited from the a_1 orbital bonding c axis pairs into the e_π orbitals bonding basal-plane neighbors. This model is consistent with the increase in the observed conductivity of Ti_2O_3 between 200 and 700°C. Between the compositions $TiO_{1.75}$–$TiO_{1.90}$, there exists a considerable number of discrete intermediate phases corresponding to the series $Ti_nO_{2n-1}(10>n>4)$. These are called shear, or Magnéli, phases.

V_2O_3 can best be synthesized by the careful reduction of V_2O_5 in a stream of pure hydrogen. The reduction is first carried out at 600°C for 4 hours and then at 1000°C for an additional 6 hours. V_2O_3, at room temperature, has the $\alpha - Al_2O_3$ corundum structure (space group $R\bar{3}c$) and exhibits metallic conductivity. Below 158 K, the symmetry becomes monoclinic with semiconducting properties (4). The corundum structure is shown in Figs. 7.16 and 7.17. The room temperature cell parameters are $a = 4.951(2)$ Å and $c = 14.004(2)$ Å. Below the transition temperature of 158 K, the resulting X-ray pattern can be indexed on the basis of a monoclinic unit cell with $a = 7.272(2)$ Å, $b = 5.000(2)$ Å, $c = 5.532(2)$ Å and $\beta = 96.65°$ (space group $I2/a$) (94,95). Dernier and Marezio (95) showed, by means of a structural refinement of the low-temperature phase, that there was an increase in V–V distances below the transition temperature. The V–V distance across a shared octahedral face increases from 2.70 to 2.745 Å, while the V-V distance across the shared octahedral edge increases from 2.872 to 2.987 Å. The displacement of edge-sharing vanadium atoms involves their movement toward the empty octahedral sites (Fig. 7.19). The observed semiconducting behavior (4) has been attributed to this expansion along the basal plane.

The one-electron energy diagrams for V_2O_3 above and below 158 K are shown in Fig. 7.20. The V(III) $3d$ orbitals are split by crystalline fields into two e_σ orbitals directed toward neighboring anions, two e_π orbitals lying in the basal plane, and one a_1 orbital directed along the c axis, as was the case for Ti_2O_3 (see Fig. 7.18). Covalent mixing of cationic $e_\sigma^2 sp^3$ and anionic sp^3 orbitals results in the formation of bonding σ and antibonding σ^* bands. Cation–cation interactions transform the a_1 orbital into bonding and antibonding bands and the e_π orbitals into a narrow band. Above 158 K, the structure of V_2O_3 is rhombohedral, the e_π band is partially filled for V(III) $3d^2$, and metallic behavior is observed. Below 158 K, the structure of V_2O_3 is monoclinic. Goodenough (42) proposed that the translational symmetry of this phase splits the e_π band into bonding and antibonding bands arising from the e_π orbital directed toward near neighbor cations, and a non-bonding state arising from the e_π orbital directed toward vacant octahedral sites. As shown in Fig. 7.20, the bonding e_π band lies below the Fermi level and is filled; the nonbonding and antibonding e_π states lie above it and are empty. Thus monoclinic V_2O_3 is a semiconductor.

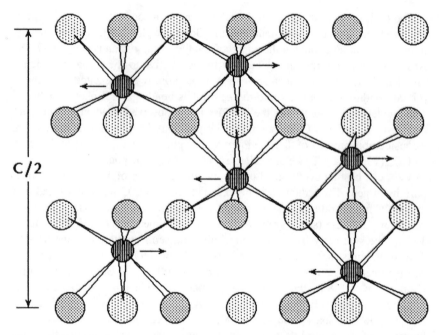

Figure 7.19. Displacement of vanadium ions in the low-temperature monoclinic phase of V_2O_3.

The electronic properties of V_2O_3 have been studied by many laboratories. It was Morin (4) who first observed the first-order electrical transition from a metal to semiconductor at 158 K on cooling. The importance of careful chemistry for the synthesis of vanadium oxide is clearly seen in the case of V_2O_3. The presence of small amounts of V(IV) has been shown by Ueda et al. (96) and Shivashankar and Honig (97), among others, to affect markedly the electronic transition observed in V_2O_3. Single crystals of V_2O_3 grown with HCl exhibited a first-order electrical transition from a metal to a semiconductor at 158 K on cooling, in agreement with the results first reported by Morin (4). However, single crystals that were prepared by chemical vapor transport using $TeCl_4$ as the transport agent remained metallic to 96 K on cooling (Fig. 7.21). The suppression of the semiconducting phase had previously been reported by Pouchard and Launay (98) and was attributed to the oxidation of V(III) to V(IV) by Cl_2.

In an attempt to verify that the V_2O_3 was being oxidized by Cl_2, Gray et al. (99) grew V_2O_3 single crystals with $TeCl_4$ and an excess of chlorine. The resistivities of these crystals were measured and they remained metallic to 77 K. To determine if the oxidized crystals contained V_3O_5, a single crystal was mounted on a Gandolfi camera and X-rayed with $CrK\alpha$ radiation. X-Ray diffraction data from this crystal could be partially indexed on the basis of a V_2O_3 unit cell; however, additional

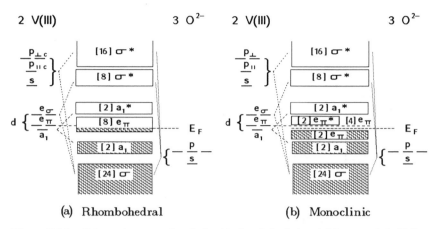

(a) Rhombohedral   (b) Monoclinic

Figure 7.20. Schematic energy bands for (a) rhombohedral and (b) monoclinic V_2O_3.

reflections appeared which could be accounted for by the presence of V_3O_5. These additional reflections were seen when the c axis of the crystal was positioned perpendicular to the X-ray beam.

Since the low-temperature transition for V_2O_3 can be observed magnetically (100), polycrystalline samples of V_2O_3 containing VO_2 were prepared and their molar susceptibilities were obtained as a function of temperature. These results are plotted in Fig. 7.22. It can be seen that with increasing V(IV) content the magnitude and temperature of the transition decrease. On the addition of 0.04 mol of VO_2, the transition for V_2O_3 has disappeared. This is in agreement with the results obtained by Ueda et al. (96), Shivashankar and Honig (97), and others. The susceptibility continually increases for V_2O_3 with the addition of VO_2. X-Ray diffraction patterns taken with a Debye-Scherrer camera were obtained for these polycrystalline samples of V_2O_3 containing V(IV). For the sample with 0.04 mol of VO_2, the monoclinic reflections associated with V_3O_5 were visible.

It can be concluded that oxidized single crystals of V_2O_3 contain V_3O_5, and that the structural similarities that exist between these two compounds allow the V_3O_5 to become an integral part of the corundum structure. The results reported in this study support the crystallographic data of Åsbrink concerning the valency distribution for V_3O_5 . The structural refinement of V_3O_5 by Åsbrink (101) indicates that the V(IV) resides in the corundum-like units of V_3O_5 while the V(III) is situated in the rutile-like units. On substitution of V(IV) ($3d^1$) for V(III) ($3d^2$), the temperature and magnitude of the transition for V_2O_3 are lowered. The attempted substitution of Ti(IV) ($3d^0$) in the corundum structure results in a decrease in the magnitude but not in the temperature of the transition.

The synthesis and crystallographic properties of the compounds belonging to the Rh–O system have been the subject of numerous studies (102–110). It has been found that three rhodium(III) oxides can exist under different conditions:

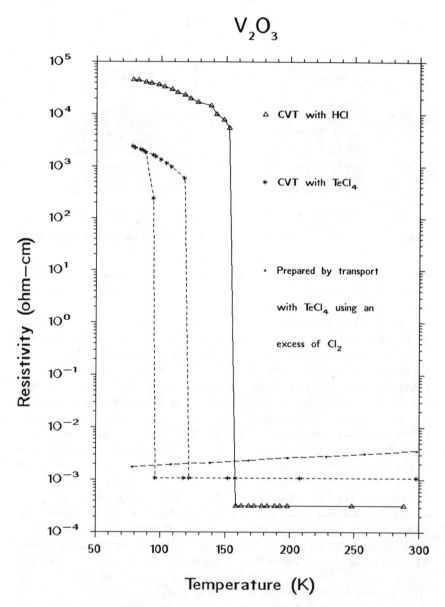

V₂O₃

V_2O_3

- △ CVT with HCl

- * CVT with TeCl₄

- • Prepared by transport with TeCl₄ using an excess of Cl₂

Resistivity (ohm–cm)

Temperature (K)

Figure 7.21. Resistivity vs temperature for various crystals of V_2O_3.

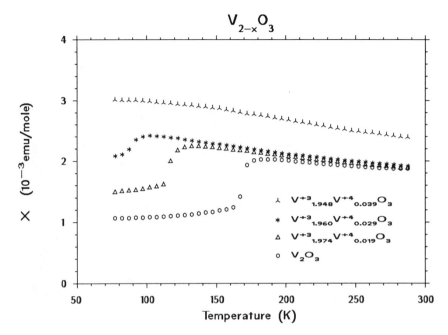

Figure 7.22. Magnetic susceptibility vs temperature for polycrystalline samples of $V_{2-x}O_3$.

the low-temperature, ambient pressure form ($Rh_2O_3(I)$); a high-temperature, high-pressure structure ($Rh_2O_3(II)$); and a high-temperature, ambient pressure compound ($Rh_2O_3(III)$).

$Rh_2O_3(I)$ has been obtained when rhodium compounds such as the hydrous oxide, nitrate, sulfate, or chloride are decomposed between 600–750°C (102–104). The resulting product crystallizes with the corundum structure, and when this phase is heated between 750 and 1000°C, it transforms irreversibly to $Rh_2O_3(III)$, which is more thermally stable. The phase $Rh_2O_3(III)$ has also been prepared by direct oxidation of rhodium metal (103,106). A major difficulty in obtaining pure samples of $Rh_2O_3(III)$ is its tendency to dissociate at relatively low temperature. The decomposition temperature depends on the oxygen pressure, and decreases from 1130°C at 760 torr to 900°C at 10 torr (104). In a reducing atmosphere, this oxide decomposes readily to rhodium metal at about 100–150°C (104). The high-temperature form $Rh_2O_3(III)$ crystallizes with an orthorhombic corundum-related structure (space group *Pbca*) (106). $Rh_2O_3(II)$ has been obtained by heating rhodium oxide (produced by heating rhodium(III) chloride in air at 800°C) at 1200–1500°C and 65 kbar pressure (107).

The crystal structure of $Rh_2O_3(III)$ was first determined by Biesterbos and Hornstra (106). The structure projected along the orthorhombic *x* axis is shown

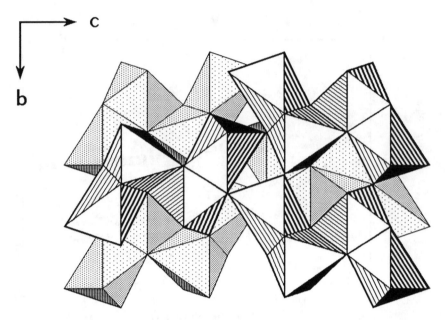

Figure 7.23. The unit cell in the structure of $Rh_2O_3(III)$ as viewed along the orthorhombic *x*-axis.

in Fig. 7.23. All of the rhodium ions are octahedrally coordinated by oxygen ions, which in turn are bonded to four rhodium ions. It can be seen that pairs of $[RhO_6]$ octahedra share faces, and each pair is connected to another pair within the plane, and to pairs above and below the plane, by means of edge sharing. The cell constants of $Rh_2O_3(III)$ were determined (110) to be $a = 5.146(3)$ Å, $b = 5.440(1)$ Å, and $c = 14.71(1)$ Å.

The sharing of RhO_6 octahedra via faces or edges allows for the possibility of direct rhodium-rhodium or indirect rhodium-oxygen-rhodium interactions which appear to determine the electrical and magnetic properties of the oxide (110). $Rh_2O_3(III)$ shows Pauli paramagnetism, and its electrical behavior is characterized by a small activation energy. The Seebeck coefficient indicates p-type conduction. This behavior is consistent with the formation of a filled *d*-band, with rhodium or rhodium-oxygen character, and acceptor levels lying close to this band. These levels are due to the presence of a small amount of Rh(IV) in the $Rh_2O_3(III)$. The determination of the magnetic as well as some of the electrical properties of $Rh_2O_3(III)$ indicates that this oxide may be considered as a semimetal (110).

References

1. A. F. Wells, *Structural Inorganic Chemistry*, 5th ed. Clarendon Press, Oxford, 1984, Chap. 12.

2. D. Watanabe, J. R. Castles, A Jostons, and A. S. Malin, *Acta Crystallogr.*, 23, 307 (1967).

3. H. Terauchi and J. B. Cohen, *Acta. Crystallogr.*, **A34**, 556 (1978).

4. F. J. Morin, *Phys. Rev. Lett.*, 3, 34 (1959); *Bell Systems Technol. J.*. 37, 1047 (1958).

5. J. B. Goodenough, *Phys. Rev.*, **117** (6), 1442 (1960).

6. J. K. Hulm, C. K. Jones, R. Mazelsky, R. A. Hein, and J. W. Gibson, *Proc. 9th Int. Conf. Low Temp. Phys.* 600 (1965).

7. P. V. Geld, S. I. Alyamovskii, and I. I. Malveenko, *Fiz. Metal. Metalloved.*, **9**, 315 (1960).

8. P. V. Geld, S. I. Alyamovskii, and I. I. Malveenko, *Phys. Metals Metallog (USSR)* **9**(2), 141 (1960).

9. P. V. Geld, S. I. Alyamovskii, and I. I. Malveenko, *Zhur. Strukt. Khim.*, 2, 301 (1961).

10. P. V. Geld, S. I. Alyamovskii, and I. I. Malveenko, *Zhur. Strukt. Khim.*, 2, 286 (1961).

11. G. Andersson, *Acta. Chem. Scand.*, **8**, 1599 (1954).

12. S. Westman and C. Nordmark, *Acta. Chem. Scand.*, **14**, 465 (1960).

13. C. G. Shull and J. S. Smart, *Phys. Rev.*, **76**, 1256 (1949).

14. L. S. Darken and R. W. Gurry, *J. Am. Chem. Soc.*, **67**, 1398 (1945).

15. E. R. Jette and F. Foote, *J. Chem. Phys.*, **1**, 29 (1933).

16. W. L. Roth, *Acta Crystallogr.*, **13**, 140 (1960).

17. J. Smuts, *J. Iron St. Inst.*, **204**, 237 (1966).

18. F. Koch and J. B. Cohen, *Acta. Crystallogr.* **B25**, 275 (1969).

19. *Strukturbericht*, **1**, 123 (1913-1928); **2**, 222 (1928-1932).

20. M. J. Redman and E. G. Steward, *Nature (London)*, **193**, 867 (1962).

21. F. Trombe and H. LaBlanchetais, *J. Phys. Rad.*, **12**, 170 (1951).

22. N. C. Tombs and H. P. Rooksby, *Nature (London)*, **165**, 442 (1950).

23. S. Greenwald and J. S. Smart, *Nature (London)*, **166**, 523 (1950).

24. J. S. Smart and S. Greenwald, *Phys.Rev.*, **82**, 113 (1951).

25. S. Greenwald, *Acta Crystallogr.*, **6**, 396 (1953).

26. S. Asbrink and L-J. Norrby, *Acta Crystallogr.*, **B26**, 8 (1970).

27. N. Datta and J. W. Jeffery, *Acta Crystallogr.*, **B34**, 22 (1978).

28. M. O'Keefe and F. S. Stone, *J. Phys. Chem. Sol.*, **23**, 261,(1962).

29. B. Roden, E. Braun, and A. Freimuth, *Solid State Commun.*, **64**,(7), 1051 (1987).

30. B. X. Yang, J. M. Tranquada, and G. Shirane, *Phys. Rev. B*, **38**,(1), 174 (1988).

31. J. B. Goodenough, *J. Mat. Ed.*, **9**(6), 619 (1987).

32. J. B. Goodenough, A. Manthiram, Y. Dai, and A. Campion. *Superconductor Sci. Technol.*, **3**, 26 (1990).

33. F. P. Koffeyberg and F. A. Benko, *J. Appl. Phys.* **53**(2) 1173, (1982).

34. L. Végard, *Phil. Mag.*, **32**, 65 (1916).

35. G. Brauer, *Handbook of Preparative Inorganic. Chemistry.*, 2nd ed. Academic Press, New York, 1965.

36. A. Wold, unpublished research.

37. A. Magnéli and G. Andersson, *Acta Chem. Scand.*, **9**, 1378 (1955).

38. B. O. Marinder and A. Magnéli, *Acta Chem. Scand.*, **12**, 1345 (1958).

39. B. O. Marinder and A. Magnéli, *Acta Chem. Scand.*, **11**, 1635 (1957).

40. B. O. Marinder, F. Dorm, and M. Seleborg, *Acta Chem. Scand.*, **16**, 293 (1962).

41. J. B. Goodenough, in *Magnetism and the Chemical Bond*, F. A. Cotton, ed. Interscience, John Wiley, New York, 1963.

42. J. B. Goodenough, *Bull. Soc. Chim. France*, **4**, 1200 (1965).

43. D. B. Rogers, R. D. Shannon, A. W. Sleight, and J. L. Gillson, *J. Inorg. Chem.*, **8**(4), 841 (1969).

44. S. N. Subbarao, Y. H. Yun, R. Kershaw, K. Dwight, and A. Wold, *Mat. Res. Bull.*, **13**, 1461 (1978).

45. L. A. Harris and R. H. Wilson, *J. Electrochem. Soc.*, **123**, 1010 (1976).

46. L. A. Harris, D. R. Cross, and M. E. Gerstner, *J. Electrochem. Soc.*, **124**, 839 (1977).

47. S. N. Subbarao, Y. H. Yun, R. Kershaw, K. Dwight, and A. Wold, *Inorg. Chem.*, **18**, 488 (1979).

48. S. Andersson, *Acta Chem. Scand.* **14**, 1161 (1960).

49. A. D. Wadsley, in *Non-Stoichiometric Compounds*, L. Mandelcorn, ed. Academic Press, New York, 1964.

50. G. Hägg, *Z. Phys. Chem.*, **B29**, 192 (1935).

51. M. L. F. Bayard, T. G. Reynolds, M. Vlasse, H. L. McKinzie, R. J. Arnott, and A. Wold, *J. Solid State Chem.*, **3**, 484 (1971).

52. J. W. Pierce, H. L. McKinzie, M. Vlasse, and A. Wold, *J. Solid State Chem.*, **1**, 332 (1970).

53. M. Nygren and M. Israelsson, *Mat. Res. Bull.*, **4**, 881 (1969).

54. S. M. Marcus and S. R. Butler, *Phys. Lett.*, **26A**, 518 (1968).

55. R. D. Shannon, *Solid State Commun.*, **6**(3) 139 (1968).

56. G. Friedheim and M. K. Hoffman, *Ber. Dtsch. Chem. Bas.*, **35**, 792 (1902).

57. O. Glemser and H. Sauer, *Z. Anorg. Allg. Chem.*, **252**, 160 (1943).

58. P. Gibart, *Compt. Rend.*, **261**, 1525 (1965).

59. J. B. MacChesney and H. J. Guggenheim, *J. Phys. Chem. Solids*, **30**, 225 (1969).

60. A. F. Wells, *Structural Inorganic Chemistry*, 5th ed. Oxford University Press, Oxford, 1984.

61. B. O. Marinder, *Arkiv. Kemi.*, **19**, 435 (1962)

62. R. C. Garvie, R. H. Hannink, and R. T. Pascoe, *Nature (London)*, **258**, 703 (1975).

63. K. Nassau, *Lapidary*, **31**, 900 (1977).

64. S. Davison, R. Kershaw, K. Dwight, and A. Wold, *J. Solid State Chem.*, **73**, 47 (1988).

65. P. Wu, R. Kershaw, K. Dwight, and A. Wold, *Mat. Res. Bull.*, 23, 475 (1988).

66. K. E. Smith, R. Kershaw, K. Dwight, and A. Wold, *Mat. Res. Bull.*, **22**, 1125 (1987).

67. B. Alexander and J. Bugosh, U.S. Patent 2,984,618, May 16, 1961.

68. E. C. Subbarao, H. S. Maiti, and K. K. Srivastava, *Phys. Stat. Sol.(A)* **21**, 9 (1974).

69. O. Ruff and F. Ebert, *Zeit. Anorg. Allgem. Chem.*, **180**, 19 (1929).

70. J. M. Marder, T. E. Mitchell, and H. Heuer, *Acta. Metall.* **31** (3), 387 (1983).

71. G. Teufer, *Acta Crystallogr.*, **15**, 1187 (1962).

72. D. K. Smith and H. W. Newkirk, *Acta Crystallogr.*, **18**, 983 (1965).

73. G. M. Wolten, *Acta Crystallogr.*, **17**, 763 (1964).

74. G. L. Clark and D. H. Reynolds, *Ind. Eng. Chem. Res.*, **29**, 711 (1937).

75. Y. Murase and E. Kato, *J. Am. Ceram Soc.*, **66**(3), 196 (1983).

76. B. H. Davis, *J. Am. Ceram. Soc.*, **67**(8), 168 (1984).

77. A. Clearfield, *Inorg. Chem.*, **3**, 146 (1964).

78. R. C. Garvie, J. Phys. Chem., *69*, 4 (1965).

79. K. V. S. R. Apparao, N. K. Sahoo, and T. C. Bagchi, *Thin Solid Films*, 129, L71 (1985).

80. M. Balog, M. Schieber, S. Patai, and M. Michman, *J. Cryst. Growth*, **17**, 298 (1972).

81. H. Sankur, J. DeNatale, W. Gunning, and J.G.Nelson, *J. Vac. Sci. Technol.*, **A5**(5), 2869 (1987).

82. D. R. McKenzie, D. J. H. Cockayne, M. G. Sceats, P. J. Martin, W. G. Sainty, and R. P. Netterfield, *J. Mat. Sci.*, **22**(10), 3725 (1987).

83. R. N. Tauber, A. C. Dumbri, and R. E. Caffrey, *J. Electrochem. Soc.*, **118**, 747 (1971).

84. M. Balog, M. Schieber, M. Michman, and S. Patai, *Thin Solid Films*, **47**, 109 (1977).

85. Y-M. Gao, P. Wu, R. Kershaw, K. Dwight, and A. Wold, *Mat. Res. Bull.*, **25**, 871 (1990).

86. J. Yahra and H. P. R. Frederikse, *Phys. Rev.*, **123**, 1257 (1961).

87. J. M. Honig and T. B. Reed, *Phys. Rev.*, **174**(3), 1020 (1968).

88. A. D. Pearson, *J. Phys. Chem. Solids*, **5**, 316 (1958).

89. L. J. Eckert and R. C. Bradt, *J. Appl. Phys.*, **44**(8), 3470 (1973).

90. C. E. Rice and W. R. Robinson, *Acta Crystallogr.*, **B33**, 1342 (1977).

91. T. B. Reed, R. E. Fahey, and J. M. Honig, *Mat. Res. Bull.*, **2**, 561 (1967).

92. L. L. VanZandt, J. M. Honig, and J. B. Goodenough, *J. Appl. Phys.*, **39**(2), 594 (1968).

93. J. B. Goodenough, *Czech. J. Phys.* **B17**, 304 (1967).

94. R. E. Newnham and Y. M. deHaan, *Z. Krist.*, **117**, 235 (1962).

95. P. D. Dernier and M. Marezio, *Phys. Rev. B*, **2**(9), 3771 (1970).

96. Y. Ueda, K. Kosuge, and S. Kachi, *J. Solid State Chem.*, **31**, 171 (1980).

97. S. A. Shivashankar and J. M. Honig, *Phys. Rev. B*, **28**(10), 5695 (1983).

98. M. Pouchard and J. Launay, *Mat. Res. Bull.*, **8**, 95 (1973).

99. M. L. Gray, R. Kershaw, W. Croft, K. Dwight, and A. Wold, *J. Solid State Chem.*, **62**, 57 (1986).

100. P. H. Carr, and S. Foner, *J. Appl. Phys. Suppl.*, **31**, 3445 (1960).

101. S. Asbrink, *Acta Crystallogr.*, **B36**, 1332 (1980).

102. G. Lunde, *Z. Anorg. Chem.* **163**, 345 (1927).

103. A. Wold, R. J. Arnott, and W. J. Croft, *Inorg. Chem.*, **2**(5), 972 (1963).

104. G. Bayer and H. G. Wiedemann, *Thermochim. Acta*, **15**, 213 (1976).

105. J. M. D. Coey, *Acta Crystallogr.*, **B26**, 1876 (1970).

106. J. W. M. Biesterbos and J. Hornstra, *J. Less Comm. Met.*, **30**, 121 (1973).

107. R. D. Shannon and C. T. Prewitt, *J. Solid State Chem.*, **2**, 134 (1970).

108. K. R. Poeppelmeier and G. B. Ansell, *J. Cryst. Growth*, **51**, 587 (1981).

109. O. Muller and R. Roy, *J. Less Comm. Metals*, **16**, 129 (1968).

110. H. Leiva, R. Kershaw, K. Dwight, and A. Wold, *Mat. Res. Bull.*, **17**, 1539 (1982).

8

Ternary and Higher Oxides

A. Synthesis and Characterization of ABO_2 Delafossite Compounds

$CuFeO_2$ is found in nature as the mineral delafossite. Early investigations utilizing X-ray diffraction, neutron diffraction, and Mössbauer techniques established the structure of $CuFeO_2$ and confirmed that Cu(I) and Fe(III) were the species present rather than Cu(II) and Fe(II) (1–5). However, the results of research carried out at E. I. du Pont deNemours in the early 1970s extended these earlier studies to include delafossite compounds containing palladium, platinum, and rhodium (6–8). These compounds were among the earliest known that contained platinum or palladium in combinations with other transition metals. In addition, $PtCoO_2$, $PdCoO_2$, $PdCrO_2$, and $PdRhO_2$ were the first compounds containing platinum or palladium with 2-fold, linear coordination by anions. The formal valence state of Pt(I) and Pd(I) was demonstrated from magnetic susceptibility measurements of $PtCoO_2$, $PdCoO_2$, $PdCrO_2$, and $PdRhO_2$. The synthesis of compounds with the delafossite structure has been achieved by several techniques summarized by Shannon et al. (6). Hydrothermal synthesis of $PtCoO_2$ was carried out in platinum or gold tubes 5 in. long by 3/8 in. in diameter under an external pressure of 3000 atmospheres and a temperature of 700°C. Platinum tubes, Co_3O_4, and a 20% aqueous hydrochloric acid solution were used to produce thin hexagonal plates of $PtCoO_2$.

The metathetical reaction between $PdCl_2$ and CoO was used to prepare $PdCoO_2$

$$PdCl_2 + 2CoO \rightarrow PdCoO_2 + CoCl_2$$

A metathetical reaction has also been used to prepare $CuFeO_2$ ($CuCl$ + $LiFeO_2$ $\rightarrow$ $CuFeO_2$ + $LiCl$).

Several of the delafossites have also been synthesized by direct solid state reactions. The synthesis of $CuFeO_2$ was achieved by reacting copper(I) oxide

(Cu$_2$O) with iron(III) oxide (Fe$_2$O$_3$) in sealed silica tubes at ambient pressure at 1000°C

$$Cu_2O + Fe_2O_3 \rightarrow 2CuFeO_2$$

An oxidizing flux reaction is an unusual method, which utilizes a low melting oxidizing flux. For the preparation of AgCoO$_2$, AgRhO$_2$, and AgCrO$_2$, a flux mixture of AgNO$_3$ and KNO$_3$ was used to promote the reaction

$$AgNO_3(\ell) + LiMO_2(s) \rightarrow LiNO_3(\ell) + AgMO_2(s) \ (M = Co, Rh, Cr)$$

The AgNO$_3$ was, therefore, one of the reactants as well as a component of the flux.

Delafossite crystallizes with the space group $R\bar{3}m$ and the structure is shown in Fig. 8.1. For PtCoO$_2$, it can be seen that the platinum and cobalt atoms occupy alternate layers in the structure. The platinum atoms are linearly coordinated to two oxygen atoms. However, the platinum atoms can also be pictured as being in hexagonal bipyramidal coordination if the surrounding platinum metal atoms

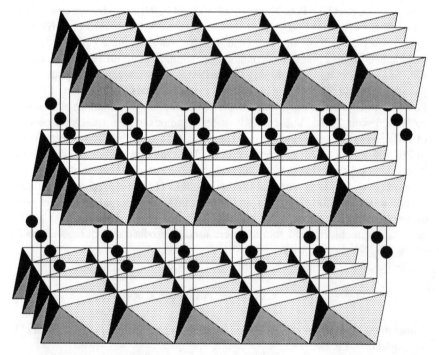

Figure 8.1. The delafossite-type structure. The solid black circles represent, for example, the platinum atoms in PtCoO$_2$.

are also considered as part of the coordination polyhedron. Each platinum atom is in a hexagonal bipyramid with respect to oxygen atoms at the apices and platinum atoms at the six equatorial positions. The cobalt atoms are in octahedral positions with respect to the oxygen atoms.

The electrical conductivities of $PdCoO_2$ and $PtCoO_2$ are anisotropic, being highest in the plane containing the a axis. This is consistent with the Pd–Pd and Pt–Pt distances of 2.83 and 2.83 Å, which are close to the metal–metal distances of 2.75 and 2.77 Å, respectively, measured in the f.c.c. metals. On the other hand, $CuFeO_2$ and $AgFeO_2$ are semiconductors, and this is consistent with the relatively large Cu–Cu distance (2.56 Å) and Ag–Ag distance (2.89 Å). Rogers et al. (8) proposed a one-electron energy diagram to rationalize these differences in the observed electronic properties of the platinum metal and non-platinum metal delafossites.

B. Perovskites

There are several excellent reviews that summarize the interesting chemistry displayed by mixed oxides crystallizing with the perovskite structure (9–11). This treatment will deal with the properties of a few compounds that presented unique chemical challenges to these writers. There are many mixed oxides, with the composition ABO_3, that are perovskites because the sum of A and B charges (+6) can be formed from (1 + 5), (2 + 4), or (3 + 3) cations. In addition, a large number of oxides with the composition $AB_{0.5}B'_{0.5}O_3$ also adopt this structure. The ideal perovskite structure, shown in Fig. 8.2, is cubic with the A species coordinated by 12 oxygen atoms and B surrounded by 6 oxygen atoms. Most ABO_3 compounds, including the mineral perovskite ($CaTiO_3$), crystallize with structures that are distorted from the ideal cubic perovskite. An example of an ideal cubic perovskite is $BaTiO_3$. Most perovskites are isostructural with $GdFeO_3$, space group *Pbnm* (D_{2h}^{16}), with four distorted perovskite units in the true crystallographic cell; six perovskites crystallize with the rhombohedral $R\bar{3}m$ (D_{3d}^5) cell with two formula units per cell. Geller and co-workers have presented a complete picture of the structural relationships existing among these compounds (12–16). A transition from the orthorhombic (*Pbnm*) to the rhombohedral structure was reported by Geller (16) to occur for $SmAlO_3$. Hence, the orthorhombic structure is linked to the ideal cubic via the rhombohedral structure. Geller concluded that the oxygen ion positions do not change toward the ideal structure, as expected from an examination of lattice constants. These anions are more favorably disposed toward transition to the rhombohedral structure. Very few of the rhombohedral crystals were expected to transform to the cubic form at any temperature.

The synthesis of perovskites containing rare earth elements has been achieved by a number of different techniques and these will now be outlined. The rare earth cobalt oxides were prepared (17) by heating mixtures of cobalt(II) carbonate

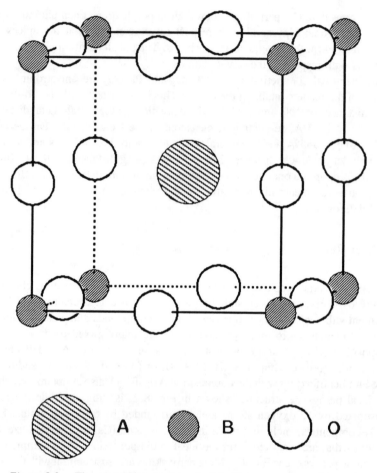

Figure 8.2. The perovskite structure (metal on corners, oxygen at center of edges) and R.E. center of cube.

and rare earth oxides or mixtures of the nitrates in air at 900°C. The purest compounds were obtained by using an excess of rare earth oxide and removing the unreacted rare earth oxide from the product with a hot saturated solution of ammonium chloride. Gallagher (18) reported the preparation of the rare earth orthoferrites and the analogous cobalt compounds by the thermal decomposition of the appropriate rare earth ferricyanide or cobalticyanide compound, e.g., $LaFe(CN)_6 \cdot xH_2O$. These compounds precipitate from aqueous solution with the appropriate rare earth to transition metal ratio, i.e., 1:1. Because the mixing is on an atomic scale, the desired compound is obtained without long and laborious grinding of reactants. For the orthoferrites, the ferrocyanide compounds $NH_4LnFe(CN)_6 \cdot xH_2O$ can be used as starting materials. The mixed rare earth cobalt oxides were also prepared (19) under high pressure (930°C, 60 kbar). $LaCoO_3$ obtained by this method is rhombohedral, space group $R\bar{3}C$ (D_{3d}^6), but the other rare earth cobalt oxides (Nd–Lu) were reported to be isostructural with $GdFeO_3$, space group *Pbnm* (D_{2h}^{16}). The rare earth vanadium oxides were prepared (17) by heating vanadium(III) oxide and the rare earth oxide in sealed silica capsules at 1200°C under vacuum. The reaction proceeded more rapidly if the mixed oxides were pressed into pellets before being sealed in the evacuated silica tubes. Under these conditions, the lanthanum, neodymium, praseodymium, and samarium compounds were obtained. To prepare lanthanum nickel(III) oxide (20), stoichiometric amounts of lanthanum and nickel(II) oxides, needed to give $LaNiO_3$, were ground together in an agate mortar with sodium carbonate (weight of Na_2CO_3 = weight of rare earth oxide). The mixture was heated in a gold crucible at 800°C and the product extracted with hot water. The employment of a sodium carbonate flux makes it possible to use lower temperatures. Solutions containing 500 ppm of dissolved $LaNiO_3$ contained only 3 ppm of sodium. $LaNiO_3$ also belongs to space group $R\bar{3}m$ with two formula weights per unit cell. When lanthanum oxide and nickel(II) oxide (La:Ni = 2:1) are heated to 1350°C, the product formed is La_2NiO_4, which crystallizes with the K_2NiF_4 structure (21). Pure $LaMnO_3$ was first prepared (21) by reacting lanthanum oxide with manganese(III) oxide under a pure nitrogen atmosphere at 1300°C. It was essential to remove all traces of oxygen from the gas stream to avoid the introduction of manganese(IV) into the product.

The precise crystal structure analysis of the ternary rare earth perovskites could not have been achieved by Geller and co-workers (12–16) without the single crystal growth achievements of Remeika (22). Equimolar proportions of the constituent oxides were transferred to a pure platinum crucible. Lead monoxide was added in the ratio of constituent oxides to lead monoxide = 1 to 6 by weight. The reactants were heated in air to 1300°C for 1 hour and uniformly cooled at 30°/hour to 850°C, then allowed to cool more rapidly to room temperature. The flux (PbO) was extracted with a hot dilute solution of nitric acid. By this method, crystals of $LnMO_3$ (M = Fe^{3+}, Al^{3+}, Sc^{3+}, Ga^{3+}, Co^{3+}) were obtained. For the orthochromites $LnCrO_3$, bismuth(III) oxide was used as the flux.

A number of the rare earth and yttrium orthoferrites $LnFeO_3$ where Ln = Y, Yb, Ho, Tb have been grown (23) under hydrothermal conditions from 20 M KOH solution at 375°C with a Δt of 30°C. A (100) seed orientation was used and the growth rate was approximately 6 mil/day.

A number of the rare earth transition metals that crystallize with the perovskite structure are ideal for the study of indirect magnetic exchange mechanisms (23–26). To interpret the magnetic data for a number of substituted manganese perovskite systems, Goodenough et al. (26) proposed several rules for the magnetic interactions present. If the cations on opposite sides of the anion (superexchange) each have a half-filled e_g orbital, maximum stability is achieved if the cations are coupled antiferromagnetically through the pair of electrons present in the anion p orbital. Where the cation on opposite sides of the anion have empty e_g orbitals, a weak antiferromagnetic coupling results. When the cations on opposite sides of the anion have empty e_g and half-filled e_g orbitals, respectively, simultaneous stabilization is achieved by ferromagnetic order.

The orthorhombic distortion from cubic symmetry observed for pure LaMnO$_3$ is larger than that for the other orthorhombic perovskites. The distinguishing feature of this distortion is that $c/\sqrt{2} < a < b$, whereas the compounds isostructural with GdFeO$_3$ have $a < c/\sqrt{2} < b$. Both structures can be indexed with space group $Pbnm/$ (D_{2h}^{16}). To distinquish between these two types of orthorhombic perovskites, Goodenough et al. (26) designated them by the symbols O' and O, respectively. The larger distortion observed in pure LaMnO$_3$ is due to superposition of a cooperative electronic ordering of the Mn^{3+} empty $d_{x^2-y^2}$ orbitals, shown in Fig. 8.3, on the ionic size distortion responsible for the O structure. This ordering results in the formation of $3d^4$ cations with half-filled e_g orbitals along the z axis and empty e_g orbitals directed along the x and y axes. Therefore, the Mn^{3+} − Mn^{3+} interactions within the c-plane are ferromagnetic; those connecting adjacent planes are antiferromagnetic. The net tetragonal distortion that occurs as a result of this ordering is sufficient to transform a lattice with orthorhombic symmetry O ($a<c/\sqrt{2}<b$) to O' ($c/\sqrt{2}<a<b$). Wold and Arnott (21) demon-

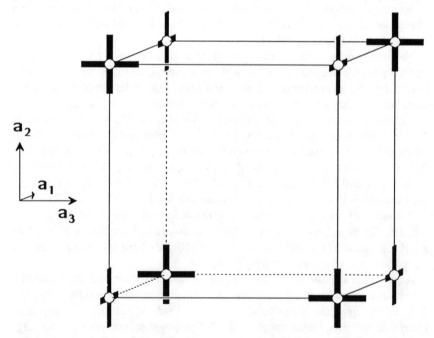

Figure 8.3. The simple cubic array of transition element ions in a perovskite-type lattice showing the orientations of empty $d_{x^2-y^2}$ orbitals responsible for orthorhombic symmetry.

strated experimentally the relative influence of the ionic-size and ligand-field effects in $LaMnO_{3+\lambda}$.

An examination (27) of the system $(La_{1-x}M_x^{2+})MnO_3$ shows that the amount of Mn^{4+} required to eliminate the cooperative distortion decreases as the size of the A cation is increased, or the magnitude of the O "pucker" decreases. This indicates that the O′ structure is the result of a superposition of electron-ordering and ionic-size distortions. Empirically, rhombohedral symmetry seems to inhibit cooperative Jahn–Teller distortions (27).

The introduction of ions such as Mn^{4+}, Cr^{3+}, Co^{3+}, and Ti^{4+}, which have empty d_{z^2} and $d_{x^2-y^2}$ orbitals on octahedral sites, reduces the number of electrons that are available to participate in the distorting mechanism. Consequently, the magnitude of the distortion and the temperature of this transition from orthorhombic to rhombohedral symmetry are expected to decrease with increasing concentration of foreign ions, as found experimentally (21).

Mn^{3+}, with an electron configuration of $3d^4$, shows a large Jahn–Teller localized distortion to tetragonal symmetry. Cooperative distortions transform the structure O to O′. When the energy involved is large enough to override the thermal energy of lattice vibrations, a long range order of the tetragonal distortion can occur, which induces an ordering of the $d_{x^2-y^2}$ orbitals (called effect A in Ref. 26). This cooperative electronic ordering results in ferromagnetic $Mn^{3+}-Mn^{3+}$ interactions within the c-plane and antiferromagnetic $Mn^{3+}-Mn^{3+}$ interactions connecting adjacent planes.

Moreover, even when the Jahn–Teller energy is insufficient to override the lattice vibrations, it will significantly favor those vibrational modes of the oxygen ions that preserve tetragonal symmetry around each cation site. As the oxygen ions vibrate, the tetragonal c axis will shift back and forth, subject only to the condition that nearest neighbors cannot have parallel c axes. Thus part of the time the instantaneous electron configuration of any two nearest neighbors will be that of empty e_g orbitals on opposite sides of the anion; the rest of the time it will have a configuration corresponding to one empty and one half-filled e_g orbital on opposite sides of the anion. Since the latter interaction is stronger than the former, the time-average interaction between any nearest neighbor pair will be ferromagnetic (effect B in Ref. 26). If only one of the two interacting cations is $3d^4$, then there can be no synchronization of orbital occupancies. The Mn^{3+} orbital involved in the interaction is partially occupied on a time-average basis, and the net interaction corresponds to having the Mn^{3+} orbital occupied because of the relative strength of the possible interactions (effect C in Ref. 26).

The introduction of sufficient ions to destroy the O′ structure of pure $LaMnO_3$ gives rise to ferromagnetism. Low-spin-state Co^{3+} and Ti^{4+} are diamagnetic, and the ferromagnetism arises from the $Mn^{3+}-Mn^{3+}$ interactions (effect B). In the system $(La,Ca)MnO_3$ there exists a rhombohedral ferromagnetic phase over the range of compositions 21-50 at % Mn^{4+}. In this range, $Mn^{3+}-Mn^{4+}$ interactions, as well as the $Mn^{3+}-Mn^{3+}$, are ferromagnetic (effects B, C), and antiferromagnetic $Mn^{4+}-Mn^{4+}$ interactions do not occur. Here, electrostatic forces maintain short-range correlation between the Mn^{4+} ions, so that Mn^{4+} ions are rarely nearest neighbors, and only electronic order is involved.

Examination of the system $La(Mn,Cr)O_3$, where Cr^{3+} ($3d^3$) has an outer-electron configuration similar to that of Mn^{4+}, gives additional confirmation of these conclusions. The $Mn^{3+}-Cr^{3+}$ and $Mn^{3+}-Mn^{3+}$ interactions are ferromagnetic if there is no e_g electron ordering in Mn^{3+} (effects B, C). The antiferromagnetic $Cr^{3+}-Cr^{3+}$ interactions, which are randomly present throughout the material, give rise to ferrimagnetism without long-range magnetic order (28).

It is interesting to note that, in contrast to these ferromagnetic or ferrimagnetic systems, $La(Mn,Fe)O_3$ remains antiferromagnetic. Since Fe^{3+} has the electronic configuration $t_{2g}^3 e_g^2$, it would be expected to couple antiferromagnetically both with Mn^{3+} and with other Fe^{3+} ions. Any tendency to form a ferromagnetic couple would be offset by the stronger antiferromagnetic interactions.

Parasitic ferromagnetism has been observed in perovskites. Early studies (29) of their magnetic properties established the fact that the magnetization varied for high fields (H>6000 Oe) according to the expression $\sigma = \sigma_o + \chi H$. The constant term represents the parasitic ferromagnetism, and χ is the susceptibility. The cause of parasitic ferromagnetism in the perovskites has not yet been established. Bozorth (30) suggested that the parasitic ferromagnetism is caused by small deflections of the B-cation moments from a uniaxial direction. This was first proposed by Dzialoshinskii (31) for some antiferromagnetic fluorides. These deflections could be induced in the orthorhombic perovskites by the crystalline fields associated with the puckering of the orthorhombic structure. From a detailed knowledge of that structure, it is possible to write down the direction of the parasitic magnetization to be associated with a given type of magnetic order and spin axis for the nearly antiferromagnetic B-cation sublattice. At low temperatures, interactions between paramagnetic rare earth cations and the Fe^{3+} sublattice may induce significant parasitic magnetization. Néel (32) pointed out that preferential ordering of lattice defects or impurities into one of the magnetic sublattices of an antiferromagnet can give rise to a weak ferrimagnetism.

Wold and Menyuk (33) observed that the parasitic magnetization found in $LaFeO_3$ can be reduced by two orders of magnitude by careful preparation of stoichiometric, more completely reacted materials. Evidently, extreme purity of the rare earth constituent is necessary before any significant measurements can be made.

C. Tungsten Bronzes and Related Compounds

a. Rhenium(VI) Oxide and Rhenium Bronzes

The ReO_3 structure is related to that of perovskite. Rhenium cations are located on octahedral sites; however, the 12-coordinated cation present in the perovskite structure is not present in ReO_3. Hence, the perovskite structure can be derived from ReO_3 by the addition of a 12-coordinated cation at the body-centered position of the unit cell.

Biltz and co-workers (34,35) first reported the preparation of ReO_3. Improved methods for its preparation were described by Nechankin et al. (36). The oxide is red and reported to be diamagnetic (37). Ferretti et al. (38) indicated that ReO_3 exists with a considerable range of composition. Since the electrical properties are dependent on the sample stoichiometry, pure stoichiometric single crystals were grown by chemical vapor transport of ReO_3 obtained by the reduction of Re_2O_7 with CO. Non-stoichiometric ReO_3 was produced by the reduction process, but purified when transported in a two-zone furnace with iodine as the transport agent. The transported, purified crystals were found to be stoichiometric within the limits of the analytical procedure used ($\pm 0.2\%$) (39). Resistivity measurements made on single crystals of stoichiometric ReO_3 (40) indicated that they exhibited metallic behavior.

In the ReO_3 structure (Fig. 8.4), the oxygen atoms occupy three-quarters of the positions in a cubic close-packed type lattice. The rhenium atoms occupy one-quarter of the available octahedral sites. The unit of structure, as represented in Fig. 8.4, is a simple cube of rhenium atoms with oxygen atoms on all of the 12 edge centers. The cubic ReO_3 structure is the simplest three-dimensional structure formed from vertex sharing of octahedral (ReO_6) groups. Each rhenium, therefore, has six oxygen nearest neighbors and each oxygen atom has two rhenium neighbors arranged linearly.

According to Morin (40) and Goodenough (41), for ReO_3 where rhenium is in the $+6$ oxidation state, the crystal environment permits the formation of a conduction band by π overlapping of the metal d orbitals with the p_π orbital of the oxygen. For the ReO_3 structure, each anion contributes two σ orbitals (sp) and two π orbitals p_π^2 (Fig. 8.5). Each octahedrally coordinated cation contributes six σ orbitals ($e_g^2sp^3$) and three π orbitals (t_{2g}^3). A large overlap of the metal t_{2g}^3 and anion p_π orbitals can result in the formation of bonding and antibonding π bands (Fig. 8.5) (41). In Fig. 8.5, the σ and π bands are composed of bonding orbitals and the σ^* and π^* bands of antibonding orbitals. The p_π^* energy level is a nonbonding state and is composed of anion p_π orbitals. The number of states is designated by [n] and refers to the spin and orbital degeneracies per molecule. ReO_3 has 25 outer electrons per molecule and hence can fill up to one-sixth of the total number of π^* states. Hence, this model is consistent with the observed metallic behavior of ReO_3.

Ferretti et al. (38) compared the electrical conductivity of single crystals of ReO_3 with that of sintered bars of polycrystalline Sr_2MgReO_6. The latter compound is an ordered perovskite and was first prepared by Longo and Ward (42). In this compound, the rhenium and magnesium atoms are at the centers of oxygen octahedra, which are linked by vertex sharing. Sr_2MgReO_6 differs from ReO_3 in that Mg(II) and Re(VI) are in a nearly perfect ordered arrangement; the rhenium

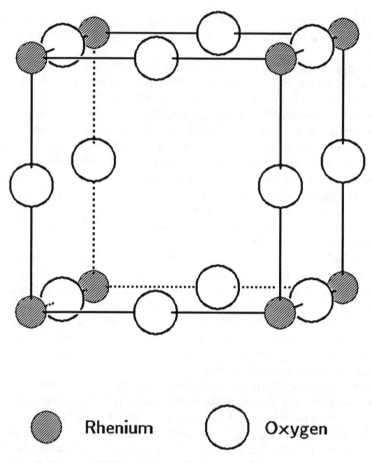

Rhenium   Oxygen

Figure 8.4. The rhenium trioxide (ReO_3) structure

atoms cannot approach each other through oxygen more closely than about 7.6 Å. However, in ReO_3, the Re atoms are about 3.8 Å apart.

Ferretti et al. (38) carried out their investigation to distinguish between models for the conduction band involving either direct overlap of t_{2g} orbitals across the face diagonal of the unit cell or indirect overlap (bonding) of the t_{2g} metal orbitals and the oxygen p_π orbital. The former model would allow both compounds to be metallic, but the latter predicts metallic behavior (delocalized d-electrons) for ReO_3 and semiconducting behavior, at most, for Sr_2MgReO_6. Sr_2MgReO_6 was found to be a semiconductor, which is consistent with metal–oxygen–metal overlap. Rhenium trioxide single crystals were found to be metallic conductors with conductivities approaching that of metallic silver. The compound was found to be diamagnetic, indicating that the weak Pauli paramagnetic moment expected

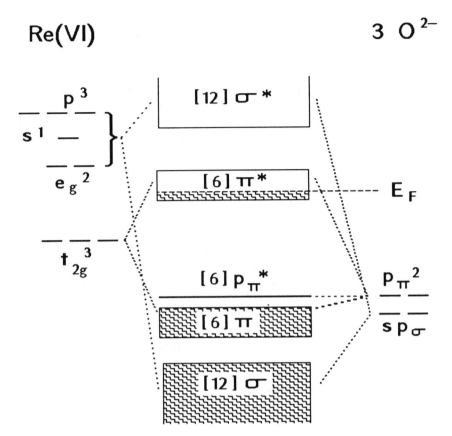

Figure 8.5. Schematic energy bands for ReO_3

from the conduction electrons is insufficient to overcome the diamagnetism of the electrons in closed shells.

b. Rhenium Bronzes

The conduction mechanism for ReO_3-type compounds, therefore, involves a strong covalent mixing between the metal t_{2g} and the oxygen p_{π} orbitals. When the resulting band is partially occupied, as in ReO_3, metallic behavior is observed. The substitution of fluorine for oxygen would increase the electron concentration in this band. However, such substitution may also decrease the degree of delocalization of the electrons, since fluorine is more electronegative than oxygen. The existence of several unstable, hygroscopic rhenium oxyfluorides such as $ReOF_4$, $ReOF_3$, ReO_3F, and ReO_2F_2 has been described by several investigators (43,44), but the system $ReO_{3-x}F_x$ has not been reported. An attempt was made (45) to prepare members of this system by hydrothermal synthesis; however, no appreciable fluorine substitution was observed when aqueous solvents were employed. Pintchovski et al. (46) developed a new synthetic technique that involves the application of hydrostatic pressure to molten salts. Ammonium hydrogen fluoride was chosen as the molten salt because it is a low melting solid (125°C), a strong fluorinating agent, and a nonaqueous solvent. The reaction of ammonium perrhenate and rhenium metal in the presence of this compound yielded a new phase with the composition $NH_4ReO_{1.5}F_3 \cdot H_2O$. Lustrous black octahedral crystals (0.05–0.50 mm on edge) of this compound were obtained. The compound crystallized in the cubic system, space group $Fm3m$, with $a_0 = 16.563$ Å. The structure consists of isolated $Re_8O_{12}F_{24}$ units bridged by ammonium ions and body-centered rhombic dodecahedra of water molecules. A plot of χ^{-1} vs T indicates that the susceptibility is temperature dependent between 77 and 800 K. Although the complexity of the spin-orbit coupling precludes an interpretation of the susceptibility data, it is clear that unpaired electrons are present in the material. The temperature dependent paramagnetic behavior, as well as the high resistivity ($\rho = 10^{10}$ Ω-cm at 298 K), indicates that $NH_4ReO_{1.5}F_3 \cdot H_2O$ is an insulator. Evidently, there are limited regions of electron delocalization (e.g., in the subunits $Re_8O_{12}F_{24}$), but there are no long-range interactions between these units. This is not surprising considering the isolation of each $Re_8O_{12}F_{24}$ unit, with a rhenium-to-rhenium distance of 6.55(2) Å.

The only other report of rhenium bronzes to date involves the use of ultrahigh pressure apparatus. Originally introduced by Bridgman (47), this technique has been varied and improved in recent years (48). Sleight and Gillson (49) report the preparation of some alkali–rhenium bronzes and solid solutions of ReO_3–WO_3 in a high-pressure apparatus described by Prewitt and Young (50). All preparations were made under a pressure of 65 kbar. The reaction mixture for the alkali bronzes was the alkali perrhenate and rhenium metal, with excess perrhenate present to act as a flux. The preparations were carried out in platinum containers at temperatures of 800–1100°C. These bronzes were bright gold in color, with an alkali content close to 0.6 (in M_xReO_3). They are highly resistant to attack by nonoxidizing acids.

The ReO_3–WO_3 solid solutions were cubic above about 25 mol% ReO_3. When prepared by directly heating the two oxides at 65 kbar and temperatures from 1000 to 1300°C, they appeared to have lost oxygen. Sleight and Gillson (49) found that the addition of Re_2O_7 prevented this oxygen loss and also acted as a flux, yielding crystals suitable for electrical measurements. A new hexagonal copper-colored rhenium oxide was discovered. Similar hexagonal bronzes were found in the sodium-rhenium and lithium-rhenium systems. All appear to be bronzes from their appearance. Table 8.1 summarizes all the phases reported by Sleight and Gillson.

c. Tungsten Bronzes

The tungsten "bronzes" were discovered by Wöhler in 1824. Whereas the original group of compounds consisted of ternary oxides of tungsten and an alkali metal,

Table 8.1. Selected Data for Some New Rhenium Oxides

Compound	Cell dimensions, (Å)			Resistivity at 25°C (Ω-cm)
	a	b	c	
ReO_3	3.7477			3.6×10^{-5}
$Na_{\approx0.6}ReO_3$	3.8304			3.0×10^{-5}
$K_{\approx0.6}ReO_3$	3.8952			7.0×10^{-5}
$ReO_3 \cdot 3WO_3$	3.7574			3.4×10^{-3}
$ReO_{3-x}(x<1)$	3.744	3.758	3.713	7.0×10^{-5}
$ReO_{3-y}(x<y<1)$	4.837	(Hexagonal)	4.518	
	4.819		4.582	
$(Li)ReO_{3-y}$	4.849	(Hexagonal)	4.488	5.1×10^{-4}
$(Na)ReO_{3-y}$	4.835	(Hexagonal)	4.535	
WO_{3-x}	5.2270	(Tetragonal)	3.8858	

bronzes have been prepared containing the transition metals Ti, V, Nb, Ta, Mo, W, Re, Ru, and Pt. A bronze is now defined as an oxide with intense color (or black), having metallic luster and showing either semiconducting or metallic behavior. A principal characteristic of bronzes is their range of composition, which results in the transition metal exhibiting a variable formal valence.

The tungsten bronzes are a group of non-stoichiometric substances that have the general formula M_xWO_3 where $0<x<1$. M can be an alkali metal (51–56), barium (57), lead (58), thallium (59), copper (60), silver (61), or one of the rare earths (62). Hägg (51) showed that the cubic sodium tungsten bronzes ($0.32 \leq x \leq 0.93$) are members of a continuous series of solid solutions. If x were equal to one, all the tungsten atoms would be present in the formal oxidation state W(V). However, with decreasing values of x, vacant positions that are statistically distributed occur in the original sodium lattice, and a corresponding number of tungsten atoms would be oxidized to the +6 state.

The structures of the tungsten bronzes result from the formation of three-dimensional networks of WO_3 by corner sharing of WO_6 octahedra. The presence of alkali metal atoms at the center of the unit cells forms cubic perovskite-like phases (Fig. 8.6). The upper limit of x in M_xWO_3 is dependent on geometric factors. When $x = 0$, the distorted ReO_3 structure is formed.

Sienko (63) regarded the alkali metal tungsten bronzes as a solid-state defect structure in which the holes in a WO_3 network are randomly occupied by alkali metal atoms. Three proposals for the origin of the conduction have been made when $x>0.3$ and metallic conductivity is observed. The first model (64) involves Na–Na bonding via direct overlap of sodium $3p$ orbitals; the second model discusses W–W bonding via direct overlap of tungsten t_{2g} orbitals (63). Goodenough (41) proposed covalent mixing of tungsten t_{2g} and oxygen p_π orbitals, which give rise to partially occupied delocalized π^* bands. For the bronzes where x is small, the structure is distorted appreciably and there is considerable

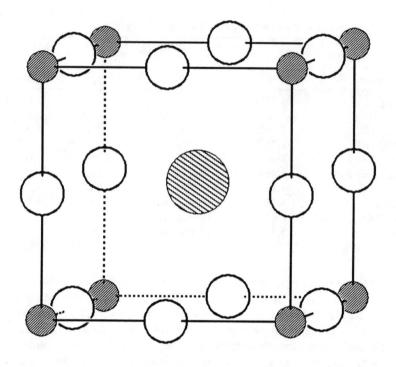

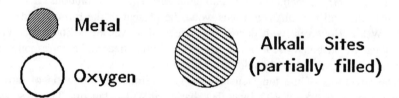

Figure 8.6. The perovskite structure of tungsten bronze

misalignment of the tungsten and oxygen atoms. This could then result in reduction in the overlap of the orbitals necessary for the formation of a conduction
band. Electrical conductivity through the crystal would then proceed by a "hopping" process, i.e., via electron transfers from lower to higher-valent ions.

There is also the possibility that in some semiconducting bronzes, e.g.,
Cu_xWO_3, the local levels are located deep in the forbidden energy gap with the
result that a finite activation energy is required for the conduction process. There
may then not be many free electrons in the conduction band, but this number
would increase with rising temperature.

The sodium tungsten bronzes Na_xWO_3 were first synthesized by Wohler (65) by the reduction of molten sodium metatungstate (Na_2WO_4) using hydrogen. Various other preparative techniques have been applied successfully to the formation of tungsten bronzes. Straumanis (66) reported preparation of cubic tungsten bronzes by the use of tungsten as a reducing agent in a mixture of tungsten(VI) oxide and an excess of sodium metatungstate. Large homogeneous crystals of Na_xWO_3 have been prepared by Ellenback and other investigators (67) by the electrolytic reduction of a fused mixture of Na_2WO_4 and WO_3 at 800°C. Electrolysis of a molten mixture by the electrolytic reduction of Li_2CO_3 and WO_3 at 780–850°C was employed by Sienko and Truong (68) to yield cubic lithium tungsten bronzes with $0.365 \leq x \leq 0.394$. Other investigators have reported the preparation of cubic $Na_{0.4}WO_3$ by sodium diffusion from single crystals of $Na_{0.678}WO_3$ packed in WO_3 powder and reacted at 920°C *in vacuo* (69).

Reactions of tungsten(VI) oxide and tungsten metal in the presence of dilute solutions of hydrofluoric acid under hydrothermal conditions resulted in the formation of a series of tetragonal oxyfluoride bronzes having the composition $WO_{3-x}F_x$ $(0.03 \leq x \leq 0.09)$ (70). These compounds are isostructural with the tetragonal bronze $Na_{0.1}WO_3$ and exhibit metallic conductivity. Examination of the system $WO_{3-x}F_x$ by Sleight (71) resulted in the preparation of two oxyfluoride phases. For $0.17 \leq x \leq 0.66$, products were obtained that gave X-ray patterns similar to those from the cubic sodium tungsten bronzes. In addition, the electrical measurements of these oxyfluorides indicated metallic behavior.

The composition $WO_{2.96}F_{0.04}$ reported by Sleight was light green in color and quite distorted from cubic symmetry. An X-ray diffraction pattern of this composition was indexed on the basis of an orthorhombic cell, whereas pure WO_3 has been reported by Magnéli (72) to be monoclinic. Although the symmetries of WO_3 and $WO_{2.96}F_{0.04}$ appear to be different at room temperature, it is possible that $WO_{2.96}F_{0.04}$ is really monoclinic with the deviation of β from 90° being too small to detect. A pressed compact of $WO_{2.96}F_{0.04}$ had a room temperature resistivity of 10^4 Ω-cm, indicating that the compound is probably not metallic (71). Thus, it appears that for the tungsten oxyfluoride bronzes, low substitutions of fluorine lead to structures that are distorted variants of the parent oxide. High substitutions lead to the formation of cubic bronze-like materials having the ReO_3 structure and exhibiting metallic behavior.

d. Molybdenum Bronzes

Molybdenum bronzes were first prepared by the electrolytic reduction of alkali metal molybdate–molybdenum(VI) oxide mixtures (73). It was indicated by Wold et al. (73) that the sodium bronzes approached the composition $NaMo_6O_{17}$. Structure analysis (74) of the phase of reported composition $Na_{0.9}Mo_6O_{17}$ indicated a monoclinic structure with space group $C2/m$, $C2$ or Cm, with cell dimensions $a = 9.57$ Å, $b = 5.50$ Å, $c = 12.95$ Å, $\beta \approx 90°$. The crystals prepared were repeatedly twinned about the normals to the (310) and (3̄10), and the composite reciprocal lattice has trigonal symmetry $P\bar{3}m1$. The sodium molybdenum bronze is a trigonally distorted perovskite structure, in which the sodium ions order in one-sixth of the voids left between the MoO_6 octahedra; it requires a randomness in the distribution of 17 oxygen atoms over 18 sites. The further distortion of the true structure to monoclinic was not fully solved by Stephenson (74), but was consistent with the ordered distribution of 340 oxygen atoms in the unit cell.

Whereas Wold et al. (73) reported the blue potassium bronze to have a composition of $K_{0.28}MoO_3$ and the red to be $K_{0.26}MoO_3$, the ideal formulas of the red and blue bronzes as worked out from X-ray crystallographic analysis (75,76) are $K_{0.33}MoO_3$ and $K_{0.30}MoO_3$, respectively. This appears to be in good agreement with the reducing power analysis for the molybdenum. The ideal compositions of $K_{0.33}MoO_3$ and $K_{0.30}MoO_3$ have been confirmed by Bouchard et al. (77).

Wold et al. (73) reported a difference in electrical conductivity between the red and blue potassium molybdenum bronzes. The red bronze ($K_{0.33}MoO_3$) showed typical semiconductor behavior with a positive temperature dependence; the blue bronze ($K_{0.30}MoO_3$) has a much lower resistivity with an apparent transition from semiconductor to metallic behavior above $-100°C$. These results were confirmed by Bouchard et al. (77). The crystal structures of the potassium molybdenum bronzes (75,76) indicate that for the red bronze the basic unit is a cluster of six distorted MoO_6 octahedra interlocked by edge sharing to give Mo_6O_{18} stoichiometry. Two K^+ ions, resulting from the ionization of two K atoms, are associated with each cluster, and two valence electrons have been incorporated in the structure. The basic unit of the blue bronze is a cluster of ten distorted MoO_6 octahedra interlocked by edge sharing to give a $Mo_{10}O_{30}$ unit. There are three K^+ ions and hence three valence electrons incorporated per cluster. In both blue and red bronzes, clusters are connected by corner sharing of octahedra through oxygens. Figure 8.7 gives a comparison of the red and blue structures based on an idealized representation using regular octahedra instead of the distorted MoO_6 groupings actually observed.

The many molybdenum bronzes that have been prepared since 1964 are classified according to color, stoichiometry, and structure. The blue bronzes have the composition $A_{0.3}MoO_3$ where A = K, Rb, and Tl; red bronzes have the stoichiometry $A_{0.33}MoO_3$ where A = Li, K, Rb, Cs, and Tl; the purple bronzes are $A_{0.9}Mo_6O_{17}$ with A = Li, Na, K, and Tl. Marcus et al. (78) reported rare earth molybdenum bronzes $A_{0.08}MoO_3$ (A = La, Ce, Eu, Gd, and Lu).

Sodium and potassium molybdenum bronzes, cubic Na_xMoO_3 and K_xMoO_3 ($x \approx 0.9$), and tetragonal K_xMoO_3 ($x \approx 0.5$) have been prepared at 65 kbar (79) and are reported to be metallic. The band model developed by Goodenough for ReO_3 and Na_xWO_3 (41,80) was first applied to the molybdenum bronzes by Dickens and Neild (81) to explain the transfer of electrons from A cations to empty π^* bands that are formed from an overlap of oxygen p_π orbitals and molybdenum t_{2g} orbitals. A recent review by Greenblatt (82) discusses the electronic and structural

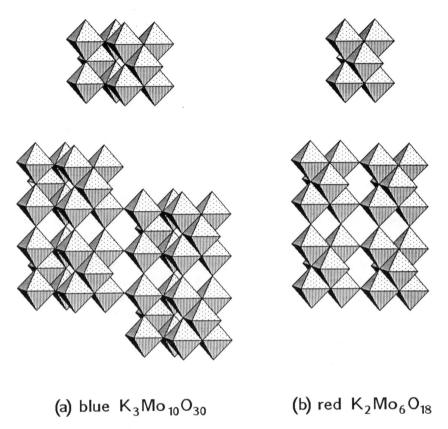

(a) blue $K_3Mo_{10}O_{30}$ (b) red $K_2Mo_6O_{18}$

Figure 8.7. Idealized packing of octahedra in two potassim molybdenum bronzes: (a) blue $K_3Mo_{10}O_{30}$ (Graham and Wadsley) and (b) red $K_2Mo_6O_{18}$ (Stephenson and Wadsley).

properties of a number of unusual molybdenum bronzes. The reader is referred to this paper, which attempts to correlate synthesis and structure with the quasi-low-dimensional properties of these compounds.

e. Oxyfluoride Bronzes of Molybdenum

Preparation of oxyfluorides in the MoO_3 system by Pierce et al. (83) has led to the formation of two new phases. The $Mo_4O_{11.2}F_{0.8}$ phase is orthorhombic and exhibits semiconductor behavior, while the $MoO_{2.4}F_{0.6}$ compound has a cubic ReO_3 type structure (space group $Pm3m$) similar to that of the tungsten bronzes.

The electrical properties of these materials can be related to differences between their structures. For the ReO_3 structure, in which the octahedra are regular and share corners in three dimensions, the conditions are optimized for the formation of collective-electron π^* orbitals of the type suggested by Goodenough (41,80).

Table 8.2. Formal Oxidation State and Coordination Number of Molybdenum in Compounds with Stoichiometries Approaching MoX_3

Compound	Oxidation state	Coordination number
MoO_3	6.00	6
$Mo_4O_{11.2}F_{0.8}$	5.80	5
$Mo_{18}O_{52}$	5.78	5
Mo_4O_{11}	5.50	6
$MoO_{2.4}F_{0.6}$	5.40	6

However, in the orthorhombic MoO_3-related structures, corner sharing occurs only along the a-axis. This is consistent with the observation that $MoO_{2.4}F_{0.6}$ is metallic whereas in $Mo_4O_{11.2}F_{0.8}$ conductivity parallel to the c-axis requires an activation energy. It can also be seen from Table 8.2 that the properties of the oxyfluorides are consistent with the prediction of Magnéli (51), namely, that as the amount of reduced molybdenum increases, the coordination number of molybdenum approaches a maximum. From Table 8.2 it can be seen that for the oxyfluoride $MoO_{2.4}F_{0.6}$ the coordination number of molybdenum is six.

The independent work of Sleight (61) confirmed the findings of Pierce. Sleight prepared, at 700°C and 3 kbar pressure, a series of cubic compounds with the composition $MoO_{3-x}F_x(0.74 \leq x \leq 0.97)$. In addition, he reported the formation of a unique compound for $x = 0.25$. This phase $Mo_4O_{11}F$ is an orthorhombic semiconductor with an activation energy of 0.13 eV (85).

f. Platinum Metal Bronzes

Waser and McClanahan (86) reported the preparation of the first platinum metal bronze $NaPt_3O_4$, which was later formulated as $Na_xPt_3O_4$. The existence of this phase was confirmed by Scheer et al. (87). The work on the platinum metal bronzes was extended by Ibers and co-workers (88). Compounds were reported with the stoichiometry $M_xPt_3O_4$ where M is an alkali or alkaline earth metal and x can vary from zero to one. With palladium as the transition metal, bronzes with Na, Ca, Cd, and Sr are known. Where platinum is the central species, the Na, Mg, and Ni bronzes have been prepared (88). Bergner and Kohlhaas (89) have also studied these compounds and investigated their thermal stabilities. The decomposition of the palladates starts at 844–1020°C and the platinates at 705–798°C. The products formed on decomposition of these bronzes are the metals and binary oxides, or more stable oxide phases.

D. Spinels

The crystal chemistry of compounds that crystallize with the spinel structure has been reviewed by Gorter (90), Hafner (91), and Blasse (92). The spinel structure takes its name from the mineral $MgAl_2O_4$. This structure, which is cubic, was first determined by Bragg (93) and Nishikawa (94). An extensive listing of the known spinels is given by Blasse (93). In the spinel structure the smallest cell having cubic symmetry contains eight formula units of the type AB_2O_4. The anions form a face-centered cubic lattice and there are 64 tetrahedral 'A' sites and 32 octahedral 'B' sites, of which only 8 and 16, respectively, are occupied. It is often convenient to distinguish the octahedral site cations by enclosing them in brackets. Two cases can be distinguished, namely, the normal spinel $A[B_2]O_4$ and the inverse spinel, $B[AB]O_4$. For the 2–3 spinels, $A^{2+}B_2^{3+}O_4$, the distribution of the cations can be

1. normal where the divalent ions are on the A sites: $A^{2+}[B^{3+}B^{3+}]O_4$

2. inverse where the divalent ions are on the B sites: $B^{3+}[A^{2+}B^{3+}]O_4$

3. intermediate between normal and inverse.

Consider, for example, the distribution of metal(II)–diiron(III) oxides. The distribution in $Zn[Fe_2]O_4$ is normal, in $Fe[NiFe]O_4$ it is inverse, and in the spinels $MnFe_2O_4$ and $MgFe_2O_4$ it is intermediate. The spinels crystallize with the space group $Fd3m$. The general features of the spinel structure are shown in Fig. 8.8, where the unit cell of edge a is subdivided into eight octants with edge $a/2$. Ideally, the anions are positioned the same way in all octants. Each octant contains four anions that form the corners of a tetrahedron. The positions of the cations are indicated in Fig. 8.8. For one octant, the occupied tetrahedral sites are the center and four corners. In the adjacent octant the central site is vacant and as a result of symmetry, half of the corner sites are occupied. The occupied octahedral B sites are located only in the latter octant. The four metal ions and the oxygen, in this octant, constitute a cube with edge $a/4$. For many spinels, the anion sublattice is distorted because of the differences in sizes of the cations. The tetrahedral sites are relatively small and generally will not provide sufficient space for the A-site cations without expanding the site. This expansion is accomplished by a displacement of the four anions away from the tetrahedral cations along the body diagonals (111 direction) of the octants containing central A-site metal ions. In the "octahedral octant," the anions are displaced in such a way that this oxygen tetrahedron shrinks by the same amount as the first expands. Thus, cubic symmetry is preserved. A quantitative measure of the anion displacement is the anion parameter u, which is shown in Fig. 8.8 and is ideally equal to 3/8. In addition to the unit cell edge, the u parameter is the only other variable distance in the spinel structure and from the u parameter, the position of every atom in the unit cell can be determined.

To consider the various interactions possible in the spinel structure, the symmetry of the cation outer-electron wave functions is important. In Fig. 8.9, it can be seen that the A-site cations have the triply degenerate t_{2g} orbitals pointing away from all neighboring ions. B-site cations have the e_g orbitals pointing toward

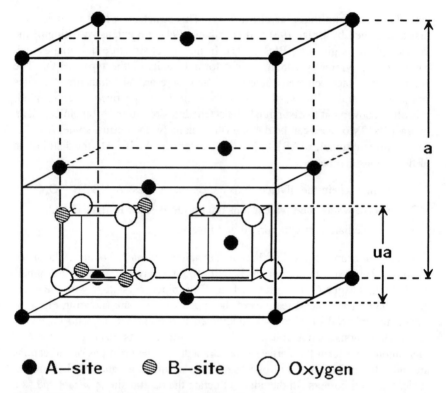

● **A–site**   ◉ **B–site**   ○ **Oxygen**

Figure 8.8. Two octants of the spinel unit cell.

near-neighbor anions, while the t_{2g} orbitals point toward neighboring B-site cations.

Each anion in the spinel structure is surrounded by one A- and three B-site cations arranged in a tetrahedron with the anion in the center. The angle AOB is about 125°, and the angle BOB is about 90°. The BB distance is $(1/4)a\sqrt{2}$, the AA distance is $(1/4)a\sqrt{3}$. The shorter BB distance is related to the fact that the anion octahedra surrounding the B-site cations share edges, whereas the anion tetrahedra surrounding the A-site cations do not have any contact.

Many compounds with the spinel structure have been reported in the literature. Most of them are oxides, although a number of spinels contain fluoride, chloride, cyanide, sulfide, selenide, and telluride ions.

It can be seen from Fig. 8.9 that the B-site cations have t_{2g} orbitals directed in such a way that they can interact with other B-site cations by direct overlap of their wave functions. For this interaction to occur, the orbitals must extend far enough out in space to overlap. The degree of direct overlap increases from nickel to titanium as the nuclear charge is reduced. This type of behavior was studied

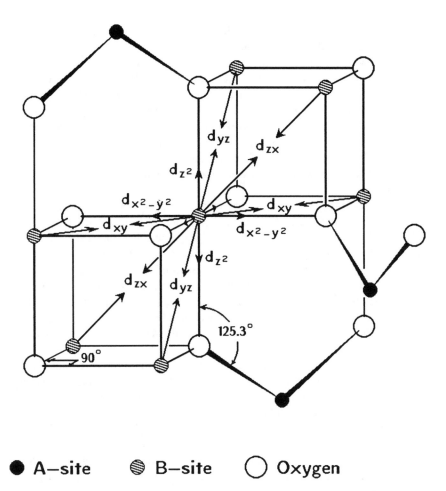

● **A–site**   ◉ **B–site**   ○ **Oxygen**

Figure 8.9.   Orientations of the B-site cation *d* orbitals with respect to the spinel structure.

by Rogers et al. (95) for a series of normal vanadium oxide spinels. If the A-site cations are decreased in size, the unit cell volume decreases and the B-site V^{3+} ions are brought closer together. Rogers et al. (95) observed that as the A-site ion decreased in size, the resistivity and activation energy decreased markedly, indicating an increase in the V^{3+}–V^{3+} overlap. It was possible by this technique to define a critical distance for overlap, R_c, below which the *d*-electrons of the transition metal ions are collective rather than localized. The concept of a critical distance was originally developed by Mott (96) and Goodenough (97–99) to explain the unusual magnetic and electrical properties of transition metal oxides where the metal ions had a $3d^n$ configuration with $n \leq 3$. For a number of oxide systems, critical distances for metal ions have been obtained that are useful in explaining changes in magnetic and electrical properties when localized electron states transform to delocalized states.

The magnetic properties of several of the spinels, in addition to their relatively low electrical conductivity compared to magnetic metals and alloys, have made them useful for a number of electronic device applications. Their intrinsic properties have led to the development of new ideas concerning the nature of magnetic exchange interactions. However, as with numerous other materials, more attention should be paid to the development of synthetic methods that result in compounds of the highest purity and homogeneity. Some of the difficulties encountered in achieving maximum purity and correct stoichiometry were discussed in Chapter 6.

The ferrites, MFe_2O_4, cannot be crystallized from aqueous solutions, although Fe_3O_4, $CoFe_2O_4$ and $MnFe_2O_4$ can be coprecipitated from solutions of soluble salts by alkalis. These materials are largely hydrous but contain the ions sufficiently ordered to exhibit spontaneous magnetization. The usual procedure employed is simply to prepare an intimate physical mixture at a temperature sufficiently high to cause the reaction of one solid with the other to form a new phase. If purity and homogeneity of the product are considered, a number of difficulties arise. It is necessary to use an increasingly refined technique in order to approach compositions corresponding exactly to the formula MFe_2O_4. It is necessary to use the purest, assayed simple oxides, accurately weighed and intimately mixed without the preferential loss of one reactant or the introduction of nonvolatile impurities. Complete homogeneity usually requires pulverizing the product and reheating two or three times. This is particularly true of ferrites prepared from refractory oxides such as MgO.

A number of variations of the direct synthesis method have been developed in which the reacting oxides are in a finely divided highly reactive state. These usually are formed by thermal decomposition of a suitable precursor. Synthetic methods of this type are discussed by Economos (100), Fresh (101), Welch (102), and Wickham et al. (103).

The iron(III) spinels melt incongruently and lose oxygen at temperatures approaching their melting points 1500-1800°C. The compounds crystallize with the cubic spinel structure and their properties were summarized by Wickham (6) and are given in Table 8.3.

Chemical transport reactions have been used (104) to prepare single crystals of iron(II) diiron(III) tetroxide (magnetite, Fe_3O_4) and other ferrites. Previously, Smiltens (105) prepared well-characterized iron(II) diiron(III) tetroxide crystals using a modified Bridgman–Stockbarger technique. The crystals were grown under a CO_2 atmosphere and then CO was introduced and the ratio of these gases changed gradually as the temperature was varied during the cooling process. This maintained the proper oxygen pressure necessary for the preparation of stoichiometric crystals. The techniques of flame fusion (106) and hydrothermal growth (107) have also been used to grow ferrite crystals, but their quality, in

Table 8.3. Properties of Several Iron(III) Spinels

Compound	Cubic lattice constant (Å)	Magnetic transition temperature (°C)	Magnetization at 0 K (BM/mol)
$MgFe_2O_4$	8.384	440	1.1–1.4
$MnFe_2O_4$	8.512	300	4.5
$CoFe_2O_4$	8.388	520	3.7
$NiFe_2O_4$	8.388	585	2.1
$ZnFe_2O_4$	8.437	−264	0

general, has usually been poor. In addition, a cobalt ferrite crystal was grown by Ferretti et al. (108) from the melt at 1600°C under an oxygen pressure of 790 psi. Chemical analysis of a portion of the crystal gave a ferrous ion content of 1.3 percent. Kunnmann et al. (109) reported that the simple ferrite crystallizes readily from the ternary flux system $MO–Na_2O–Fe_2O_3$. Crystals of $CoFe_2O_4$ were grown from a mixture of 1.5 mol of $CoFe_2O_4$ and 1 mol of $Na_2Fe_2O_4$ utilizing a platinum wire cold finger. The crystals grown were approximately 1 in. in diameter and 3/4 in. thick. However, the crystals were not dense and appeared to have many defects. Nickel ferrite crystals were also grown by Kunnmann et al. (110) from sodium ferrite flux by a modified Czochralski method. The crystals were analyzed and corresponded to the composition $Ni_{1.01}Fe_2O_{4.01}$.

Relatively pure crystals of magnetite and the other ferrites can best be grown by the method of chemical transport. In this method (104, 111) the charge material reacts with the transport agent to form a more volatile compound. This vapor diffuses along the tube to a region of lower temperature, where some of the vapor undergoes the reverse reaction. The starting compound is reformed and the transport agent is liberated. The latter then diffuses to the hot end of the tube and again reacts with the charge. Under the proper conditions, the compound can be deposited as crystals. This can be illustrated for the growth of Fe_3O_4 crystals with HCl as the transport agent. The reversible reaction can be represented by

$$Fe_3O_4 + 8HCl \rightleftharpoons FeCl_2 + 2FeCl_3 + 4 H_2O$$

Crystals of Fe_3O_4 grown by chemical vapor transport are well-formed octahedra, with shiny faces measuring up to 5 mm on an edge. X-Ray diffraction patterns of the ground-up crystals were identical to that of the starting material. The results of chemical analysis of a typical magnetite run are given in Table 8.4. There was no evidence of chloride in the final transported crystals, although traces might be expected.

The "precursor" method has also been applied (112) to the preparation of the chromites, which have the formula MCr_2O_4 and possess the spinel structure.

Table 8.4. Results of Chemical Analysis

Fe_3O_4	Total iron	Iron (II)	Formula
Calculated	72.3670	24.12	
Found (powder)	72.66	23.78	$Fe^{2+}_{0.99}Fe^{3+}_{2.03}O_4$
Found (crystal)	72.83	23.65	$Fe^{2+}_{0.98}Fe^{3+}_{2.01}O_4$

Previous methods employed for the synthesis of chromites consisted simply of preparing an intimate physical mixture of two appropriate oxides and heating this mixture to a sufficiently high temperature (1400–1700°C) to cause the two oxides to react to form the desired product. The method does not yield pure products easily because of the refractory nature of chromium(III) oxide and of many of the divalent-metal oxides involved. Tedious grinding procedures and firings must be done, and extraneous impurities are usually introduced as a result of the grinding. Ignition at elevated temperatures occasionally results in the preferential loss of some of the divalent oxides, e.g., ZnO, CuO.

The precursor methods used by Whipple and Wold (112) to prepare a number of chromites were discussed in Chapter 6. These precursors achieve excellent stoichiometry, low trace impurity content, and homogeneity approaching the maximum theoretically possible. The lattice constants for the various chromites prepared from precursors are given in Table 8.5. They are in good agreement with those reported in the literature. The cubic to tetragonal transformation point of iron chromite was found to be $-138\pm3°C$. This point is considerably lower than $-90°C$ reported by Francombe (113). The lower value would indicate some improvement in the homogeneity and purity of the iron chromite prepared by the precursor method. The saturation magnetic moments (n_B) for a number of transi-

Table 8.5. Crystallographic and Magnetic Properties of Chromites

Compound	Cubic a_o (Å)	Tetragonal	Transformations	μ (BM/mol)[a]
$MgCr_2O_4$	8.333 ± 0.002			0.15
$NiCr_2O_4$		$a = 8.248 \pm 0.002$ $c = 8.454 \pm 0.002$	Tetragonal to cubic at $37\pm2°C$	
$MnCr_2O_4$	8.437 ± 0.002			1.20
$CoCr_2O_4$	8.332 ± 0.002			0.18
$CuCr_2O_4$		$a = 8.532 \pm 0.003$ $c = 7.788 \pm 0.003$		0.72
$ZnCr_2O_4$	8.327 ± 0.002			0.12
$FeCr_2O_4$	8.377 ± 0.002		Tetragonal to cubic at $-138\pm3°C$	0.84

[a] Measured at 10 kOe and 4.2 K.

tion metal chromites are also given in Table 8.5 (112). Lotgering (114) indicated that the chromite (oxygen) spinels have a B–B interaction that is of the same order of magnitude as the A–B interaction. This interaction usually results in lower saturation moments than are predicted by the Néel theory of ferrimagnetism.

E. Conducting Iron Tungstates and Niobates

The use of n-type semiconductors as photoanodes for the photoassisted decomposition of water using sunlight is well known (115–117). However, a few p-type semiconductor photocathodes have been reported in the literature. The $A(II)B(VI)O_4$ tungstates and molybdates are potential compounds for use as photoelectrodes because these materials have the possibility of being either p-type or n-type semiconductors. P-type behavior could be observed when a small amount of one of the transition metals is oxidized, e.g., the introduction of a small amount of iron(III) as Fe_2WO_6 in iron(II) tungstate. In addition, n-type behavior might be produced when one of the metals in the structure is reduced to a lower oxidation state, e.g., the reduction of W(VI) or Mo(VI) to W(V) or Mo(V). Thus, the ABO_4 tungstates and molybdates have potential use as either photoanodes or photocathodes in the photoassisted decomposition of water.

The divalent transition metal tungstates such as $FeWO_4$ crystallize with the wolframite structure (118). Wolframite is the name given to the mineral with composition $(Fe,Mn)WO_4$ and the structure is a variant of α-PbO_2. The structure, which is shown in Fig. 8.10, is monoclinic with $Z = 2$ and space group $P2/c$ (C_{2h}^4). For pure $FeWO_4$ (feberite), $a = 4.730$ Å, $b = 5.703$ Å, $c = 4.952$ Å, and $\beta = 90°05'$ (119). The structure consists of a hexagonal close-packed oxygen array in which one-half of the octahedral holes are occupied. The cation distribution in the octahedral interstices gives rise to zigzag chains of skew-edge linked octahedra extending along the c axis, and in any single chain there is only one type of cation. The zigzag chains are arranged in alternating layers, perpendicular to the a direction, of iron and tungsten chains. Between layers, the chains are connected by corner sharing octahedra so that no chain of one type of cation is linked to a chain of the same cation.

Single crystals and polycrystalline samples of $FeWO_4$ were prepared and characterized by Sieber et al. (120). From high temperature paramagnetic data, the presence of high spin state iron(II) $3d^6$ was confirmed. Qualitative Seebeck measurements indicated p-type conductivity, and the room temperature resistivity of single crystals was ≈ 100 Ω-cm with an activation energy of 0.16(2) eV. The p-type conductivity is consistent with a small amount of iron(III) being present due to the formation of a small quantity of Fe_2WO_6 and its ability to form a solid solution with $FeWO_4$.

Fe_2WO_6 crystallizes with the tri-α-PbO_2 structure above 800°C (space group $Pbcn$). The tri-α-PbO_2 structure, like the wolframite structure, is an ordered

Figure 8.10. The wolframite structure: *c* axis perspective of linked octahedra; light shaded octahedra represent [WO₆] units.

variant of α-PbO$_2$ (121,122). It consists of a distorted hexagonal close-packed array of oxygen anions in which one-half of the octahedral interstices are occupied by iron and tungsten cations in an ordered manner. The cations are distributed in such a way as to give rise to skew-edge linked chains of octahedra extending along the *c* direction as shown in Fig. 8.11. Separate chains are corner linked to each other. Senegas and Galy (121) indicated that ideally one-third of these puckered chains contains only iron atoms, while two-thirds show a one-to-one ordering of iron and tungsten atoms. The 2:1 cation ordering causes a tripling of the *b* axis relative to the normal α-PbO$_2$ unit cell. The space group is *Pbcn* and the lattice parameters for the orthorhombic cell are $a = 4.577(1)$ Å, $b = 16.750(1)$ Å, and $c = 4.965(1)$ Å.

Fe$_2$WO$_6$ prepared from stoichiometric mixtures of the appropriate oxides always shows trace amounts of α-Fe$_2$O$_3$ in the product. This is indicated by the presence of some of the strongest reflections of this oxide (012, 110, 024, 116, 214) near the limit of detection. This material has been reported (122) to be an n-type semiconductor with an activation energy of 0.17 eV and a room temperature resistivity of $\approx$50 Ω-cm. The presence of trace amounts of α-Fe$_2$O$_3$ can be accounted for on the basis of a solid solution of small amounts of FeWO$_4$ in Fe$_2$WO$_6$ (122). Thus, the extrinsic n-type semiconducting behavior of Fe$_2$WO$_6$ is consistent with the solid solution of FeWO$_4$ in Fe$_2$WO$_6$, thereby introducing iron(II) and iron(III) on equivalent sites so that conduction may occur by electron hopping along the chains of the tri-α-PbO$_2$ structure.

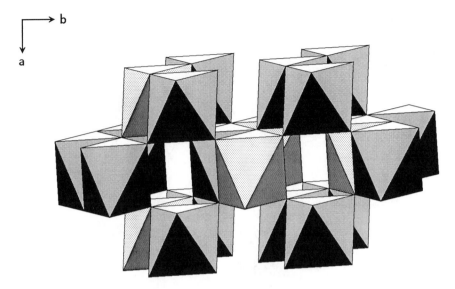

Figure 8.11. The tri$-\alpha-$PbO$_2$ structure: c axis perspective of linked octahedra; light shaded octahedra represent [WO$_6$] units.

FeNbO$_4$ has been reported to crystallize with the monoclinic wolframite structure, space group $P2/c$ (C_{2h}^4) below 1085°C. Roth and Waring (123) and Laves et al. (124) have shown that between 1085° and 1380°C, a transition to orthorhombic α-PbO$_2$, space group $Pbcn$ (D_{2h}^{14}), can occur that will further transform above 1380°C to the tetragonal rutile structure, space group $P4_2/mnm$ (D_{4h}^{14}), shown in Fig. 8.12.

It is well known (125) that if the sites occupied by M and M' in MM'X$_4$ (FeNbO$_4$) or MM'$_2$X$_6$ (FeNb$_2$O$_6$) are those occupied by M in MX$_2$, the more complex structures may be described as superstructures. Hence, in FeNbO$_4$, the octahedral sites between alternate pairs of close-packed layers in the α-PbO$_2$ structure are occupied by Fe and Nb atoms, and in the columbite structure (FeNb$_2$O$_6$), there is a more complex type of replacement. However, in all of the structures, one-half of the octahedral sites are occupied in an hexagonal close-packed array of anions.

It has also been indicated by Turnock (126) that FeNb$_2$O$_6$ may be incorporated in solid solution with FeNbO$_4$. By this mechanism, one might expect that the resistivity of FeNbO$_4$ can be adjusted by the presence of controlled small concentrations of Fe(II). The monoclinic α-PbO$_2$ phase of FeNbO$_4$ was prepared by Koenitzer et al. (127). The room temperature resistivity of sintered disks was found to be 40(1) Ω-cm. The disks were shown to be n-type by qualitative measurement of the Seebeck effect. However, routine measurements were unable to detect any Hall voltage in these compounds, which indicates that their mobilit-

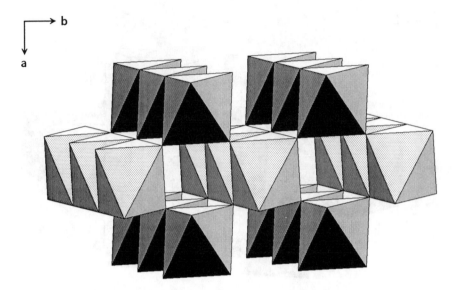

Figure 8.12. The rutile structure: c axis perspective of linked octahedra.

ies are less than 0.1 cm^2/V-sec, as would be expected for a hopping conductor (128). Given the above resistivity, the minimum carrier concentration, of the order of 2×10^{18} cm^{-3}, can then be estimated for the FeNbO$_4$ disks. It would be anticipated that the resistivity of intrinsic FeNbO$_4$ would be much higher than 40 Ω-cm. However, as Turnock indicated (126), FeNb$_2$O$_6$ can be incorporated in solid solution with FeNbO$_4$. The formation of such solid solution would be consistent with relatively high conductivity and the phase separation of a few percent of α-Fe$_2$O$_3$. Careful examination of X-ray patterns obtained from the sintered discs showed the presence of α-Fe$_2$O$_3$.

F. High-Temperature Superconducting Oxides

The discovery of superconductivity in oxides containing copper has resulted in numerous systematic investigations of these materials. A large group of layered cuprates with the general stoichiometry $(ACuO_{3-x})_m(A'O)_n$ has been prepared and characterized. These compounds show an intergrowth of multiple oxygen-deficient perovskite layers with rock salt type layers. These intergrowth structures consist of superconducting active layers containing CuO$_2$ sheets with a constant oxygen concentration and inactive layers of variable oxygen concentration. The superconductor compositions have a mixed formal valence in the CuO$_2$ sheets and occur in a narrow compositional range (129).

Many reviews have been devoted to the structural principles and nonstoichiometry phenomena in connection with the superconducting properties of these materials, but little has appeared concerning several serious chemical problems that will prevent these materials from achieving widespread commercial use. Among the most serious problems that would have to be overcome are the high mobility of copper and the extreme reactivity of holes, i.e., Cu(III). Furthermore, many of the studies up to 1991 have indicated the presence of impurity phases, which is not surprising considering the chemical complexity of the systems studied. This brief treatment will summarize a number of systems starting with those derived from the K_2NiF_4 structure type. This will be followed by a summary of some of the chemical problems associated with the $YBa_2Cu_3O_7$ (1–2–3) compounds, and finally, the more complex bismuth and thalium superconducting oxides. Despite some 6 years of intense research in many laboratories, and literally thousands of research papers, the fundamental chemical and structural problems remain to be solved.

Superconducting oxides have been known since 1964, but until recently the intermetallic compounds showed higher superconducting transition temperatures. In 1975, research scientists at E. I. du Pont de Nemours (130) discovered superconductivity in the system $BaPb_{1-x}Bi_xO_3$ with a T_c of 13 K. The structure for the superconducting compositions in this system is only slightly distorted from the ideal perovskite structure. It is generally accepted that a disproportionation of the Bi(IV) occurs, namely, $2Bi(IV) (6s^1) \rightarrow Bi(III)(6s^2) + Bi(V)(6s^0)$ at approximately 30% Bi. Sleight found that the best superconductors were single phases prepared by quenching from a rather restricted single phase region, and hence these phases are actually metastable materials. At equilibrium conditions, two phases with different values of x would exist; the phase with a lower value of x would be metallic and that with a higher value of x would be a semiconductor. It is important to keep in mind that the actual assignment of formal valence states is a convenient way of electron accounting; the actual states include appreciable admixing of anion functions. The system $BaPb_{1-x}Bi_xO_3$ should be studied further since it contains compositions showing the highest T_c for any superconductor not containing a transition element. Cava and Batlogg (131) have shown that $Ba_{0.6}K_{0.4}BiO_3$ gives a T_c of almost 30 K, which is considerably higher than the 13 K reported for $BaPb_{0.75}Bi_{0.25}O_3$.

New superconducting copper oxides have been synthesized with transition temperatures as high as 125 K ($Tl_2Ba_2Ca_2Cu_3O_{10}$). All of the known copper oxide superconductors have a common structural feature, namely the presence of $[CuO_2]_\infty$ planes. Copper is partially oxidized beyond the formal valence Cu(II) and charge compensation can be achieved in the case of La_2CuO_4 by the substitution of Sr^{2+} for La^{3+}. La_2CuO_4 has the K_2NiF_4 structure and can be considered as an intergrowth between the perovskite and rock salt structures (Fig. 8.13), $MO \cdot AMO_3 = AM_2O_4$. In this structure, there is no Cu–O–Cu bonding perpendicular to the (CuO_2) sheets. La_2CuO_4 was reported by Longo and Raccah (132) to show an orthorhombic distortion of the K_2NiF_4 structure with $a = 5.363$ Å, $b = 5.409$ Å, and $c = 13.17$ Å. It was also reported (133,134) that La_2CuO_4 has a

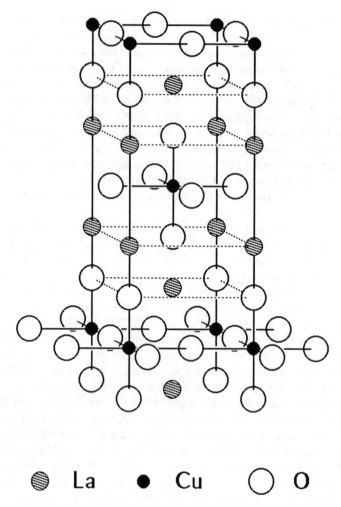

Figure 8.13. The ideal K_2NiF_4 structure of La_2CuO_4.

variable concentration of anion vacancies that may be represented as La_2CuO_{4-x}. The extent of the anion vacancies was reexamined (135) and the magnitude of this deficiency is less that can be unambiguously ascertained by direct thermogravimetric analysis, which has a limit of accuracy in x of 0.01 for the composition La_2CuO_{4-x}. However, significant shifts in the Néel temperature confirm a small variation in anion vacancy concentration.

It was shown by Mitsuda et al. (134) that in the system $La_2CuO_{4-x'}$, the Néel temperature is sensitive to the oxygen vacancy concentration. Some samples of pure La$_2$CuO$_4$ have been reported to show no antiferromagnetic ordering down to 4.2 K (133,136–138). However, DiCarlo et al. (135) found that pure La$_2$CuO$_4$ always showed a T_N. The removal of small amounts of oxygen from pure La$_2$CuO$_4$ did shift T_N from 240 to 300 K. Samples of La$_2$CuO$_{4.00(1)}$ (135) prepared at 500°C and pressures up to 2000 psi of O$_2$ still showed a Néel temperature of 240 K. All of the samples studied by DiCarlo et al. (135) also showed p-type conductivity. This is consistent with the model proposed by Goodenough (129,138) (Fig. 8.14) in which antiferromagnetic behavior gives rise to correlation splitting of the σ^* band. The observed p-type character of these samples, as well as of samples in which strontium is substituted for lanthanum, La$_{1.8}$Sr$_{0.2}$CuO$_4$, is consistent with the existence of holes in the π^* band.

In the La$_{2-x}$A$_x$CuO$_4$ phases (A = Ca, Sr, and Ba), the substitution of the alkaline earth cation for the rare earth depresses the tetragonal-to-orthorhombic transition temperature. The transition disappears completely at $x > 0.2$, which is about the composition for which superconductivity is no longer observed. Compositions of La$_{2-x}$A$_x$CuO$_4$ can also be prepared (139,140) where A is Cd(II) or Pb(II). However, these phases are not superconducting, and it appears that the more basic divalent cations are necessary to allow for the stabilization of Cu(III). The ESR spectra indicate the absence of a localized square planar Cu(II) species and the phase shows Pauli-paramagnetic behavior over the temperature range from 77 to 300 K. The Pauli-paramagnetic behavior of La$_{1.8}$Sr$_{0.2}$CuO$_4$ is consistent with delocalized electrons (141). Subramanian et al. (142) recently substituted sodium and potassium into La$_2$CuO$_4$, giving rise to the compositions La$_{2-x}$A$_x$.

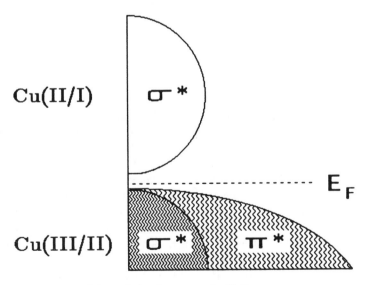

Figure 8.14. Schematic band structure for CuO.

CuO_4. However, only the sodium substituted samples exhibited superconducting behavior. The partial substitution (<10%) of Pr, Nd, Gd, or Eu for La lowers the transition temperature T_c.

The compound $YBa_2Cu_3O_7$ shows a superconducting transition at ≈ 93 K and crystallizes as a defect perovskite (Fig. 8.15). The unit cell of $YBa_2Cu_3O_7$ is orthorhombic (*Pmmm*) with $a = 3.8198(1)$ Å, $b = 3.8849(1)$ Å, and $c = 11.6762(3)$ Å. The structure may be considered as an oxygen-deficient perovskite with tripled unit cells due to Ba–Y ordering along the c axis. For $YBa_2Cu_3O_7$, the oxygens occupy 7/9ths of the anion sites. One-third of the copper is in 4-fold coordination and two-thirds are 5-fold coordinated. A reversible structural transformation occurs with changing oxygen stoichiometry $YBa_2Cu_3O_x$ going from orthorhombic at $x = 7.0$ to tetragonal at $x = 6.0$ (143). The value $x = 7.0$ is achieved by annealing in oxygen at 400–500°C, and this composition shows the sharpest superconducting transition, $T_c \approx 90$ K. It was shown by Davison et al. (141) that these materials are readily decomposed by water and carbon dioxide in air to produce carbonates.

Recently Maeda et al. (144) reported that a superconducting transition of 120 K was obtained in the Bi/Sr/Ca/Cu/O system. The structure was determined for the composition $Bi_2Sr_2CaCu_2O_8$ by several laboratories (145–147).

In most of the studies reported to date on the Bi/Sr/Ca/Cu/O system, measurements were made on single crystals selected from multiphase products. The group at du Pont selected platy crystals having the composition $Bi_2Sr_{3-x}Ca_xCu_2O_{8+y}$ (0.9>x>0.4) which showed a $T_c \approx 95$ K. Crystals of $Bi_2Sr_{3-x}Ca_xCu_2O_{8+y}$ for $x = 0.5$ gave orthorhombic cell constants of $a = 5.399$ Å, $b = 5.414$ Å, $c = 30.904$ Å (145). These cell dimensions are consistent with the results of other investigators (146,147). The structure (Fig. 8.16) consists of pairs of CuO_2 sheets interleaved by Ca(Sr) alternating with double bismuth oxide layers. Sunshine et al. (147) indicated that the addition of Pb to this system raises the T_c above 100 K. There are now three groups of superconducting oxides that contain the mixed Cu(II)–Cu(III) oxidation states, namely $La_{2-x}A_x^{2+}CuO_4$ where A^{2+} = Ca, Sr, Ba, $LnBa_2Cu_3O_7$ where Ln is almost any lanthanide, and $Bi_2Sr_{3-x}Ca_xCu_2O_{8+y}$.

Sheng and Hermann (148) have recently reported on a high-temperature superconducting phase in the system Tl/Ba/Ca/Cu/O. Two phases were identified by Hazen et al. (149), namely, $Tl_2Ba_2CaCu_2O_{8+\delta}$ and $Tl_2Ba_2Ca_2Cu_3O_{10+\delta}$. Sleight et al. (145,150) have also reported on the structure of $Tl_2Ba_2CaCu_2O_8$ as well as $Tl_2Ba_2CuO_6$. In addition, the superconductor $Tl_2Ba_2Ca_2Cu_3O_{10}$ has been prepared (151) and shows the highest T_c of any known bulk superconductor, namely ≈ 125 K. A series of oxides with high T_c values have now been studied for the type $(A^{III}O)_2A_2^{II}Ca_{n-1}Cu_nO_{2+2n}$ where A(III) is Bi or Tl, A(II) is Ba or Sr, and n is the number of Cu–O sheets stacked. To date, $n = 4$ is the maximum number of stacked Cu–O sheets examined. There appears to be a general trend whereby T_c increases as n increases. Unfortunately, these phases display rather complex ordering; crystals are grown in sealed gold tubes and excess reactants are always

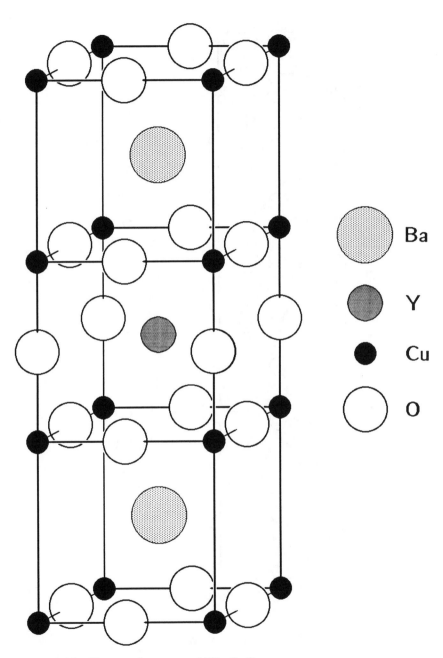

Figure 8.15. The crystal structure of $YBa_2Cu_3O_7$.

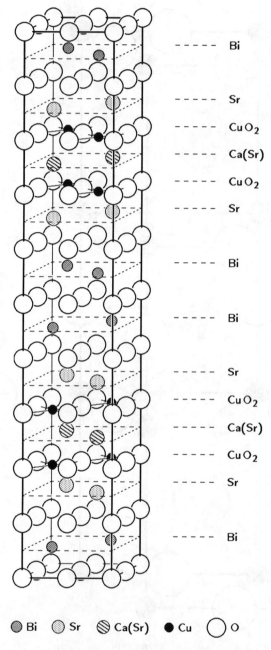

Bi

Sr

CuO_2

Ca(Sr)

CuO_2

Sr

Bi

Bi

Sr

CuO_2

Ca(Sr)

CuO_2

Sr

Bi

◉ Bi   ◯ Sr   ⊛ Ca(Sr)   ● Cu   ◯ O

Figure 8.16.   The crystal structure of
$Bi_2Sr_{3-x}Ca_xCu_2O_8$.

present. The toxicity as well as the volatility of thallium augment the problems encountered in obtaining reasonable quantities of homogeneous single-phase materials.

References

1. A. Pabst, *Am. Mineral.*, **31**, 539 (1946).

2. W. Soller and A. J. Thompson, *Phys. Rev.*, **47**, 644 (1935).

3. H. Wiedersich, J. W. Savage, A. H. Muir Jr., and D. G. Swarthout, *Mineral. Mag.*, **36**, 643 (1968).

4. A. H. Muir and H. Weidersich, *J. Phys. Chem. Sol.*, **28**, 65 (1967).

5. A. Apostolov, *God. Sofii. Univ. Fiz. Fak.*, **59**, 47 (1966).

6. R. D. Shannon, D. B. Rogers, and C. T. Prewitt, *Inorg. Chem.*, **10**(4), 713 (1971).

7. C. T. Prewitt, R. D. Shannon, and D. B. Rogers, *Inorg. Chem.*, **10**(4), 719 (1971).

8. D. B. Rogers, R. D. Shannon, C. T. Prewitt and J. L. Gillson, *Inorg. Chem.*, **10**(4), 723 (1971).

9. R. Ward, in *Progress in Inorganic Chemistry* Vol. 1. Interscience, New York, 1959.

10. F. S. Galasso, *Structure, Properties and Preparation of Pervoskite-Type Compounds*. Pergamon, Oxford, 1969.

11. J. B. Goodenough and J. M. Longo, *Crystallographic and Magnetic Properties of Perovskites and Provskite-Related Compounds*, Landolt-Börnstein Group III/Volume 4A, 1970.

12. S. Geller, *J. Chem. Phys.*, **24**, 1236 (1956).

13. S. Geller, *Acta Crystallogr.*, **10**, 248, (1957).

14. S. Geller and V. B. Bala, *Acta Crystallogr.*, **9**, 1019 (1956).

15. S. Geller and E. A. Wood, *Acta Crystallogr.*, **9**, 563 (1956).

16. S. Geller, *Acta Crystallogr.*, **10**, 243 (1957).

17. A. Wold and R. Ward, *J. Am. Chem. Soc.*, **76**, 1029 (1954).

18. P. K. Gallagher, *Mat. Res. Bull.*, **3**, 225 (1968).

19. G. Demazeau, M. Pouchard, and P. Hagenmuller, *J. Solid State Chem.*, **9**, 202 (1974).

20. A. Wold, B. Post, and E. Banks, *J. Am. Chem. Soc.*, **79**, 4911 (1957).

21. A. Wold and R. J. Arnott, *J. Phys. Chem. Solids*, **9**, 176 (1959).

22. J. P. Remeika, *J. Am. Chem. Soc.*, **78**, 4259 (1956).

23. J. B. Goodenough, *Phys. Rev.*, **100**(2), 564 (1955).

24. A. Wold, R. J. Arnott, and J. B. Goodenough, *J. Appl. Phys.*, **29**, 387 (1958).

25. A. Wold and W. Croft, *J. Phys. Chem.*, **63**, 447 (1959).

26. J. B. Goodenough, A. Wold, R. J. Arnott, and N. Menyuk, *Phys. Rev.*, **124** (2), 373 (1961).

27. G. H. Jonker, *Physica*, **22**, 707 (1956).

28. U. H. Bents, *Phys. Rev.*, **106**(2), 225 (1957).

29. R. Pauthenet and P. Blum, *C. R. Acad. Sci.*, **239**(33) (1954).

30. R. M. Bozorth, *Phys. Rev. Lett.*, **1**(10), 362 (1958).

31. I. E. Dzialoshinskii, *J. Exp. Theoret. Phys. (USSR)*, **32**, 1547 and **33**, 1454 (1957) [trans., Soviet Phys., JETP **6**, 1120 and 1259 (1957)].

32. L. Néel, *C. R. Acad. Sci.*, **239**(1), 8 (1954).

33. A. Wold and N. Menyuk, unpublished research.

34. W. Biltz, A. A. Lehrer, and K. Meisel, *Nachr. Ges. Wiss. Gottingen, Math. Phys. Klasse*, **191** (1931).

35. W. Biitz, A. A. Lehrer, and K. Meisel, *Z. Anorg. Allgem. Chem.*, **207**, 113 (1932).

36. N. Nechankin, A. D. Kurtz, and C. F. Hiskey, *J. Am. Chem. Soc.*, **73**, 2828 (1951).

37. J. H. E. Griffiths, J. Owen, and I. M. Ward, *Proc. Roy. Soc.*, **A219**, 526 (1953).

38. A. Ferretti, D. B. Rogers, and J. B. Goodenough, *J. Phys. Chem. Solids*, 26, 2007 (1965).

39. D. G. Wickham and E. R. Whipple, *Talanta*, **10**, 314 (1963).

40. F. J. Morin, *J. Appl. Phys.*, **32**, 2195 (1961).

41. J. B. Goodenough, *Bull. Soc. Chim. France, 4*, 1200 (1965).

42. J. Longo and R. Ward, *J. Am. Chem. Soc.*, **83**, 2816 (1961).

43. G. B. Hargreaves and R. D. Peacock, *J. Chem. Soc.*, 1099 (1960).

44. A. Engelbrecht and A. V. Grosse, *J. Am. Chem Soc.*, **76**, 2042 (1954).

45. E. E. Aynsley, R. D. Peacock, and P. L. Robinson, *J. Chem. Soc.*, 3376 (1958).

46. F. Pintchovski, S. Soled, R. G. Lawler, and A. Wold, *Inorg. Chem.*, **15**, 330 (1976).

47. P. W. Bridgman, *The Physics of High Pressures*, G. Bell and Sons, London, 1958.

48. R. S. Bradley, ed. *High Pressure Physics and Chemistry*, Academic Press, New York, 1963.

49. A. W. Sleight and J. L. Gillson, *Solid State Commun.*, **4**, 601 (1966).

50. C. T. Prewitt and H. S. Young, *Science*, **149**, 535 (1965).

51. G. Hägg, *Z. Phys. Chem.*, **B29**, 192 (1935).

52. F. Wohler, *Ann. Chim. Phys.*, **43**(2), 29 (1823).

53. A. S. Ribnick, B. Post, and E. Banks, *Adv. Chem. Ser.*, **39**, 246 (1963).

54. M. Atoji and R. E. Rundle, *J. Chem. Phys.*, **32**, 627 (1960).

55. F. Kupka and M. J. Sienko, *J. Chem. Phys.*, **18**, 1296 (1950).

56. A. Magnéli, *Nova Acta Regiae Soc. Sci. Upsaliensis*, **14**, 3 (1950).

57. L. E. Conroy and T. Yokokawa, *Inorg. Chem.*, **4**, 994 (1965).

58. R. A. Bernoff and L. E. Conroy, *J. Am. Chem. Soc.*, **82**, 6261 (1960).

59. M. J. Sienko, *J. Am. Chem. Soc.*, **81**, 5556 (1954).

60. L. E. Conroy and M. J. Sienko, *J. Am. Chem. Soc.*, **79**, 4048 (1957).

61. M. J. Sienko and B. R. Mazumder, *J. Am. Chem. Soc.*, **82**, 3508 (1960).

62. W. Ostertag, *Inorg. Chem.*, **5**, 758 (1966).

63. M. J. Sienko, Paper 21, *Adv. Chem. Ser.*, **39**, 224 (1963).

64. A. R. Mackintosh, *J. Chem. Phys.*, **38**(8), 1991 (1963).

65. F. Wohler, *Am. Phys.*, **2**, 350 (1824)

66. M. E. Straumanis, *J. Am. Chem. Soc.*, **71**, 679 (1949).

67. L. D. Ellenback, H. R. Shanks, P. H. Sidles, and G. C. Danielson, *J. Chem. Phys.*, **35**, 298 (1961).

68. M. J. Sienko and T. B. N. Truong, *J. Am. Chem. Soc.*, **83**, 3939 (1961).

69. W. McNeill and L. E. Conroy, *J. Chem. Phys.*, **36**, 87 (1962).

70. T. G. Reynolds and A. Wold, *J. Solid State Chem.*, **6**, 565 (1973).

71. A. W. Sleight, *Inorg. Chem.*, **8**, 1764 (1969).

72. A.S.T.M. X-ray Powder Diffraction File Card 5-0363 data supplied by A. Magnéli.

73. A. Wold, W. Kunnmann, R. J. Arnott, and A. Ferretti, *Inorg. Chem.*, **3**, 545 (1964).

74. N. C. Stephenson, *Acta Crystallogr.*, **20**, 59 (1966).

75. N. C. Stephenson and A. D. Wadsley, *Acta Crystallogr.*, **19**, 241 (1965).

76. J. Graham and A. D. Wadsley, *Acta Crystallogr.*, **20**, 93 (1966).

77. G. H. Bouchard, Jr., J. Perlstein, and M. J. Sienko, *Inorg. Chem.*, **6**, 1682 (1967).

78. J. Marcus, C. Escirbe-Fillippini, R. Chevalier, and R. Buder, *Solid State Commun.*, **62**(4), 221 (1987).

79. T. A. Bither, J. L. Gillson, and H. S. Young, *Inorg. Chem.*, **5**, 1559 (1966).

80. J. B. Goodenough, *Czech. J. Phys.*, **B17**, 304 (1967).

81. P. G. Dickens and D. J. Neild, *Trans. Faraday Soc.*, **B17**, 304 (1967).

82. M. Greenblatt, *Chem. Rev.*, **88**, 31 (1988).

83. J. W. Pierce, H. L. McKinzie, M. Vlasse, and A. Wold, *J. Solid State Chem.*, **1**, 332 (1970).

84. A. Magnéli, *J. Inorg. Nucl. Chem.*, **2**, 330 (1956).

85. B. L. Chamberland, *Mat. Res. Bull.*, **6**, 425 (1971).

86. J. Waser and E. D. McClanahan, Jr., *J. Chem. Phys.*, **19**(4), 413 (1951).

87. J. J. Scheer, A. E. Van Arkel, and B. Heyding, *Can. J. Chem.*, **33**, 683 (1955).

88. D. Cahen, J. A. Ibers, and R. D. Shannon, *Inorg. Chem.*, **11**, 2311 (1972).

89. D. Bergner and R. Kohlhaas, *Z. Anorg. Chem.*, **401**, 15 (1973).

90. E. W. Gorter, *Philips Res. Rep.*, **9**, 295 (1954).

91. S. Hafner, *Schweiz. Minerol. Petrogr. Mitt.*, **40**, 207 (1960).

92. G. Blasse, *Philips. Res. Rep.*, **S19**(3), 1 (1964).

93. W. H. Bragg, *Nature (London)*, **95**, 561 (1915), *Phil. Mag.* **30**, 305 (1915).

94. S. Nishikawa, *Proc. Tokyo-Math. Phys. Soc.*, **8**, 199 (1915).

95. D. B. Rogers, R. J. Arnott, A. Wold, and J. B. Goodenough, *J. Phys. Chem. Solids*, **24**, 347 (1963).

96. N. F. Mott, *Proc. Phys. Soc. (London)*, **A62**, 416 (1949).

97. D. G. Wickham and J. B. Goodenough, *Phys. Rev.* **115**(5), 1156 (1959).

98. J. B. Goodenough, *Magnetism and the Chemical Bond*, Interscience, New York, 1963.

99. J. B. Goodenough, *Phys. Rev.*, **117**(6), 1442 (1960).

100. G. Economos, *J. Am. Ceram. Soc.*, **38**, 241, 292 (1955).

101. D. L. Fresh, *Proc. I. R. E.*, **44**, 1303 (1956).

102. A. J. E. Welch, *Proc. Inst. Elec. Eng.*, **104**, Part B, Suppl. No. 5, 138 (1957).

103. D. G. Wickham, E. R. Whipple, and E. G. Larson, *Inorg. Nucl. Chem.*, **14**, 217 (1960).

104. Z. Hauptman, *Czech. J. Phys.*, **B12**, 148 (1962).

105. J. Smiltens, *J. Chem. Phys.*, **20**, 990 (1952).

106. R. Lappa, *Przeglad Telekohmunik Acrying*, **31**, 229 (1958).

107. N. Yu Ikornikova, *Dokl. Akad. Nauk. USSR*, **130**, 610 (1960).

108. A. Ferretti, R. J. Arnott, E. Delaney, and A. Wold, *J. Appl. Phys.*, **32**(5), 905 (1961).

109. W. Kunnmann, A. Wold, and E. Banks, *J. Appl. Phys.*, **33**(3), 1364 (1962).

110. W. Kunnmann, A. Ferretti, and A. Wold, *J. Appl. Phys.* **34**(4) (Part 2), 1264 (1963).

111. R. Kershaw and A. Wold, *Inorg. Synth.*, **11**, 10 (1968).

112. E. Whipple and A. Wold, *J. Inorg. Nucl. Chem.*, **24**, 23 (1962).

113. M. H. Francombe, *J. Phys. Chem. Solids*, **3**, 37 (1957).

114. F. K. Lotgering, *Philips Res. Rep.*, **11**, 190 (1956).

115. A. J. Nozik, *Annu. Rev. Phys. Chem.*, **29**, 189 (1978).

116. J. Manassen, G. Hodes, and D. Cahen, *Chemtech.*, **11**(2), 112 (1981).

117. D. E. Scaife, *Solar Energy*, **25**, 41 (1980).

118. R. O. Keeling, Jr., *Acta Crystallogr.*, **10**, 209 (1957).

119. D. Ulkü, *Z. Krist.*, **124**, 192 (1967).

120. K. Sieber, K. Kourtakis, R. Kershaw, K. Dwight, and A. Wold, *Mat. Res. Bull.*, **17**, 721 (1982).

121. J. Senegas and J. Galy, *J. Solid State Chem.*, **10**, 5 (1974).

122. H. Leiva, K. Dwight, and A. Wold, *J. Solid State Chem.*, **42**, 41 (1982).

123. R. S. Roth and J. L. Waring, *Am. Mineral.*, **49**, 242 (1964).

124. Von F. Laves, G. Bayer, and A. Panagos, *Schweiz. Minerol. Petrog. Mitt.*, **43**, 217 (1963).

125. A. F. Wells, *Structural Inorganic Chemistry*, 4th ed. Oxford University Press (Clarendon), London, 1975, p.48.

126. A. C. Turnock, *J. Am. Ceram. Soc.*, **49**(4), 177 (1966).

127. J. Koenitzer, B. Khazai, J. Hormadaly, R. Kershaw, K. Dwight, and A. Wold, *J. Solid State Chem.*, **35**, 128 (1980).

128. C. A. Ackert and J. Volger, *Phys. Lett.*, **8**, 244 (1964).

129. J. B. Goodenough, A. Manthiram, Y. Dai, and A. Campion, *Superconductor Sci. Technol.*, **3**, 26 (1990).

130. A. W. Sleight, J. L. Gillson, and P. E. Biestedt, *Solid State Commun.*, **17**, 27 (1975).

131. R. J. Cava and A. B. Batlogg, *Nature (London)*, **332**, 814 (1988).

132. J. M. Longo and P. M. Raccah, *J. Solid State Chem.*, **6**, 526 (1973).

133. D. C. Johnston, J. P. Stokes, D. P. Goshorn, and J. T. Lewandowski, *Phys. Rev. B*, **36**, 4007 (1987).

134. S. Mitsuda, G. Shirani, S. K. Sinha, D. C. Johnston, M. S. Alverez, D. Vaknin, and D. E. Moncton, *Phys. Rev. B*, **36**, 822 (1987).

135. J. DiCarlo, C-M. Niu, K. Dwight, and A. Wold, *Chemistry of High-Temperature Superconductors II*, ACS Symposium Series No. 377, p. 140 (1988).

136. N. Nguyen, F. Studer, and B. Raveau, *J. Phys. Chem. Solids*, **44**, 389 (1983).

137. T. Frelthoft, G. Shirane, S. Mitsuda, J. P. Remeika, and A. S. Cooper, *Phys. Rev. B*, **37**, 137 (1988).

138. J. B. Goodenough, *J. Mat. Educ.*, **9**(6), 619 (1987).

139. I. S. Shaplygin, B. G. Kakhan, R. Lazareo, and V. B. Russ., *J. Inorg. Chem.*, **24**, 820 (1979).

140. J. Gopalakrishnan, M. Subramanian, C. C. Torardi, J. P. Attfield, and A. W. Sleight, *Mat. Res. Bull.*, **24**(3), 321 (1989).

141. S. Davison, K. Smith, Y-C. Zhang, J-H. Liu, R. Kershaw, K. Dwight, P.H.Rieger, and A. Wold, *Chemistry of High-Temperature Superconductors*, Symposium Series No. 351, p. 65 (1987).

142. M. A. Subramanian, J. Gopalakrishnam, C. C. Torandi, T. R. Askew, R. B. Flypen, A. W. Sleight, J. J. Lin, and S. J. Poon, *Science*, **240**(4851), 495 (1988).

143. P. K. Gallagher, H. M. O'Bryan, S. A. Sunshine, and D. W. Murphy, *Mat. Res. Bull.*, **22**, 995 (1987).

144. H. Maeda, Y. Tanaka, M. Fukutomi, and T. Asano, *Jpn. J. Appl. Phys.*, **27**, L209 (1988).

145. M. A. Subramanian, C. C. Torardi, J. C. Calabrese, J. Gopalakrishnan, K. J.

Morrissey, J. R. Askew, R. B. Flippen, U. Chowdhry, and A. W. Sleight, *Science*, **239**, 1015 (1988).

146. J. M. Tarascon, Y. LePage, P. Barboux, B. G. Bagley, L. H. Greene, W. R. McKinnon, G. W. Hull, M. Giroud, and D. M. Hwang, *Phys. Rev. B*, **37**(16), 9382 (1988).

147. S. A. Sunshine, T. Siegrist, L. F. Schneemeyer, D. Murphy, R. J. Cava, B. Batlogg, R. B. van Dover, R. M. Fleming, S. H. Glarum, S. Nakahara, R. Farrow, J. J. Krajewski, S. M. Zahurak, J. V. Wasczak, J. H. Marshall, P. Marsh, L. W. Rupp Jr., and W. F. Peck. *Phys. Rev. B*, **38**(1), 893 (1988).

148. Z. Z. Sheng and A. M. Hermann, *Nature (London)*, **332**(6160), 138 (1988).

149. R. M. Hazen, L. W. Finger, R. J. Angel, C. T. Prewitt, N. L. Ross, C. G. Hadidiacos, P. J. Heaney, D. R. Veblen, Z. Z. Sheng, A. ElAli, and A. M. Hermann, *Phys. Rev. Lett.*, **60**(16), 1657 (1988).

150. C. C. Torardi, M. A. Subramanian, J. C. Calabrese, J. Gopalakrishnan, E. M. McCarron, K. J. Morrissey, T. R. Askew, R. B. Flippen, U. Chowdhry, and A. W. Sleight, *Phys. Rev. B*, **38**(1), 225 (1988).

151. C. C. Torardi, M. A. Subramanian, J. C. Calabrese, J. Gopalkrishnan, K. J. Morrissey, T. R. Askew, R. P. Flippen, U. Chowdhry, and A. W. Sleight, *Science*, **240**(4852), 631 (1988).

152. A. B. Batlogg, *Physica*, **26B**, 275 (1984).

Problems

P8.1. Pure WO_3 is a yellow solid that can be grown as single crystals. The compound shows high resistivity and has a band gap greater than 2 eV. On reduction with small amounts of hydrogen, the compound turns green then blue and its resistivity decreases though its band gap remains unchanged. A more severe reduction with hydrogen results in metallic conduction and ultimately metallic tungsten is the sole product.

    a. How can pure WO_3 crystals be grown?

    b. Discuss the oxide chemistry of tungsten as reduction proceeds to metallic tungsten.

    c. Why does the conductivity of WO_3 on reduction change even when only small quantities of oxygen are removed?

    d. Name two other ways in which WO_3 can be converted into a phase showing high conductivity.

P8.2. Many solids with the stoichiometry ABO_3 crystallize with the perovskite structure.

    a. Indicate how the radii of A, B, and O determine the stability of this structure and also indicate the coordination of A and B with respect to O.

    b. $BaTiO_3$ is a white solid, with a resistivity $> 10^6$ Ω-cm and a band gap of $\approx$ 3 eV. On gentle reduction in a sealed ampoule with Ti°, the compound becomes black, has the same band gap, but the resistivity is reduced to as low as $0.1\ \Omega$-cm. Indicate what has happened and account for the change in properties in terms of the electronic state of Ti.

c. What is a sodium tungsten bronze? How is it prepared? What are its electrical properties and how have they been explaned in terms of a 1-electron band model?

d. If the compounds $LaFe_{0.5}Cr_{0.5}O_3$ and $LaFe_{0.5}Co_{0.5}O_3$ could be prepared as ordered perovskites (B-site ordering), they would show spontaneous ferromagnetism. Show how this may be accounted for in terms of the simple rules of superexchange.

P8.3. If MgO and SiO_2 (Fig. P8.3) are placed in contact and heated, what compound(s) would be expected to form on the SiO_2 and MgO faces?

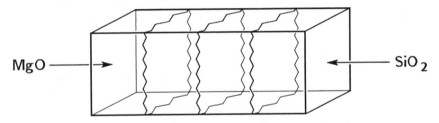

Figure P8.3

P8.4. Account for the following:

a. The oxides MnO, FeO, CoO, and NiO all have the cubic rock salt structure with octahedral coordination of the cations. The structure of CuO is different and contains grossly distorted CuO_6 octahedra.

b. ZnO when heated to 900°C turns deep yellow but turns white on cooling to room temperature.

P8.5. What are the precise structural relationships between the following pairs of compounds:

a. $MgFe_2O_4$–Fe_2SiO_4
b. Fe_2O_3–$FeTiO_3$
c. γFe_2O_3–Fe_3O_4
d. $LaNiO_3$–La_2NiO_4

P8.6. $Ba(Pb_{1-x}Bi_x)O_3$ is a mixed valence perovskite (130). Single crystals have been grown by Batlogg (152) and at $x = 0.25$ show a T_c of 11–12 K.

a. How would single crystals be grown if this compound melts incongruently?

b. Describe the perovskite structure in terms of the coordination of the metal ions by oxygen.

c. Discuss the valencies of all ions present in the structure. [Hint: One ion is present in two formal oxidation states.]

d. What problems might exist concerning metal–to–oxygen ratio, i.e., what deviations from ABO_3 might occur and how would this be determined?

e.   Describe two techniques to demonstrate the existence of a superconducting transition.

P8.7.   A number of publications have appeared describing the superconducting behavior of alkaline earth substituted superconducting La_2CuO_4. Some of these compounds have been reported to show transitions to over 40 K.

a.   How can such materials be prepared to give precise stoichiometries of desired compositions for Ca^{2+}, Ba^{2+}, and Sr^{2+} substituted phases?

b.   Why would Sr^{2+} be the best of these ions to substitute for La^{3+}?

c.   How would the valencies of the copper present and the oxygen content be ascertained?

d.   Would you expect single crystals of alkaline earth substituted La_2CuO_4 to be readily grown and, if so, by what technique. If not, why not?

P8.8.   Figure P8.8 illustrates a member of a well-known group of materials. Answer each of the questions concerning this specific compound.

a.   Give the formula of the above compound and indicate how you arrived at the stoichiometry based on the structure.

b.   What is the structure type and what parameters determine whether a compound of the proper stoichiometry crystallizes with this structure?

c.   How is the above compound prepared as a powder?

d.   Give and discuss one technique that can be used to grow single crystals of this material. The technique must be one that will successfully produce a stoichiometric single crystal.

e.   How would you measure the electrical band gap of the material?

f.   What would be the resistivity of a stoichiometric composition of the above phase and the sign of the carrier when slightly oxygen deficient?

g.   How would you increase the conductivity significantly to give properties characteristic of a metal?

h.   What electronic model would you use to explain this increase in conductivity?

i.   How could you differentiate by magnetic measurements the differences in the phases discussed in (f) and (g)?

P8.9.   a.   Pure FeO has never been prepared as a stoichiometric compound. What is the best way of approaching pure FeO?

b.   If a pure crystal of stoichiometric FeO could be prepared how would its electronic and magnetic properties differ from TiO?

P8.10.   The superconducting properties of 1–2–3 superconductors are dependent on the crystallographic and electronic structure of the copper species present. Explain this statement. What is the role of the metal-to-oxygen ratio (be specific).

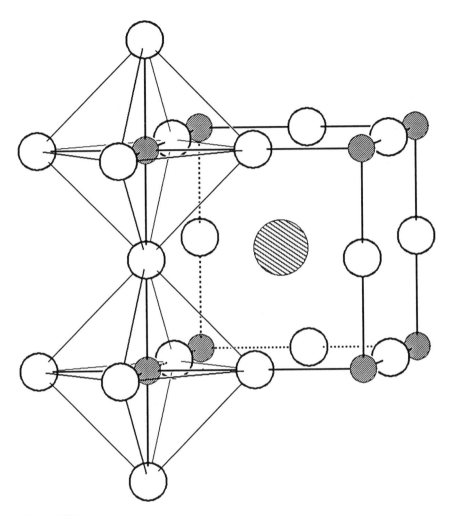

Figure P8.8

Part III
Selected Metal Sulfides

9

Binary Sulfides

A. Low-Temperature Syntheses of Transition Metal Sulfides

A major interest of solid state chemists has been the development of novel synthetic methods for the low-temperature preparation of transition metal sulfides. The traditional method used to prepare these compounds involves the direct combination of the elements in evacuated silica tubes. Complete reaction requires the use of high temperatures for long periods of time. To ensure homogeneity, it is necessary to expose periodically fresh metal surface to the chalcogen. This is accomplished by one of two methods. The sample can be removed from the tube intermittently and ground under an inert atmosphere, or it can be mechanically shaken while it is still sealed in the evacuated tube. Direct combination of the elements can result in homogeneous, single-phase materials, but the products are highly crystalline and have low surface areas. These features are undesirable for certain applications, e.g., in catalytic processes. A number of ternary compounds in which anion substitution has been achieved [e.g., CoP_xS_{2-x} (1), $CoAs_xS_{2-x}$ (2), and $CoSe_xS_{2-x}$ (3)] have also been prepared by direct combination of the elements.

Another method for the preparation of transition metal chalcogenides is by a vapor phase reaction of the metal halide with hydrogen sulfide at elevated temperatures. The reaction of $TiCl_4$ with hydrogen sulfide proceeds at 450°C as indicated by the following equation (4):

$$TiCl_4 + 2H_2S \xrightarrow{450°C} TiS_2 + 4HCl$$

The elevated temperatures are necessary for achieving favorable reaction rates. However, the products obtained by this method have low surface areas and are well crystallized.

A number of novel synthetic techniques for the low-temperature preparation of transition metal chalcogenides will now be discussed. The compounds selected are those that have been of particular interest from the point of view of potential catalyst applications or because of their unique magnetic/structural properties.

a. Low-Temperature Syntheses and Properties of Co_9S_8, Ni_3S_2, and Fe_7S_8

Delafosse et al. (5–8) have shown that sulfides of nickel and cobalt can be prepared by heating their anhydrous sulfates in a stream of H_2/H_2S at low temperatures. However, there was no report of the experimental conditions used to obtain the phases Ni_3S_2 or Co_9S_8, which are of importance as hydrodesulfurization catalysts. It has been shown (9,10) that both Co_9S_8 and Ni_3S_2 permit little variation from ideal stoichiometry.

Synthetic samples of the low-temperature phase of Fe_7S_8 have been prepared by Lotgering (11) and magnetic measurements confirmed the conclusions of other investigators (12–14) that the spontaneous magnetism of Fe_7S_8 represents a ferrimagnetic structure that is based on an ordering of iron vacancies. This can be represented by the formula:

$$Fe_4 \uparrow \; [Fe_3 \downarrow \; \Box]S_8.$$

Bertaut (13) discussed the ferrimagnetic behavior of naturally occurring Fe_7S_8 samples in terms of the ordering of iron vacancies as well as of spins. In a study by Pasquariello et al. (15), a quenched sample of Fe_7S_8 showed a temperature-independent susceptibility from liquid nitrogen to room temperature, which was considered to be consistent with a random distribution of iron vacancies. The observed magnitude of 25×10^{-6} emu/g for the susceptibility of the quenched sample was consistent with an antiferromagnet well below T_N. The annealed samples of Fe_7S_8 showed strong field-dependent behavior, i.e., large spontaneous magnetization, which coincided with the appearance of superlattice lines in the X-ray diffraction patterns. These observations were consistent with the Bertaut model for vacancy ordering in ferrimagnetic Fe_7S_8.

Co_9S_8 and Ni_3S_4 were prepared from $CoSO_4 \cdot 7H_2O$ and $NiSO_4 \cdot 6H_2O$, respectively (15). The sulfates were predried at 135°C for 4 hours and then heated first at 250°C for 1 hour in a dry nitrogen atmosphere, followed by a 2-4 hour treatment in a 40/1 (v/v) H_2/H_2S atmosphere at 525°C. The Fe_7S_8 was prepared from predried $Fe_2(SO_4)_3 \cdot nH_2O$ by heating in a 10/1 H_2/H_2S atmosphere at 325°C followed either by quenching or slow cooling the product to give either random or ordered distribution of the vacancies.

Co_9S_8 crystallizes in the space group $Fm3m$. The cubic close packing of sulfur atoms in this compound results in a cubic cell with $a = 9.923(1)$ Å (16). In Co_9S_8, cobalt atoms occupy 1/8th of the available octahedral and 1/2 of the tetrahedral sites. One out of every nine cobalt atoms is in an octahedral position, with the rest occupying pseudotetrahedral sites. Single crystals of Co_9S_8 were grown by chemical vapor transport using iodine as the transport agent (16). The crystals were determined to be cation deficient having a composition of $Co_{8.85}S_8$. Co_9S_8 was observed to be metallic and Pauli paramagnetic (15). The unit cell is shown in Fig. 9.1, and an important feature of the structure is the presence of three metal–metal bonds that extend from each tetrahedral cobalt to form cube clusters of tetrahedral cations (Fig. 9.2). The metallic behavior of Co_9S_8 is a result of both direct and indirect metal–metal interactions (16).

The low-temperature rhombohedral form of Ni_3S_2 is stable below 556°C (10). The unit cell has the symmetry of the space group $R32(D_3)$ and the dimensions

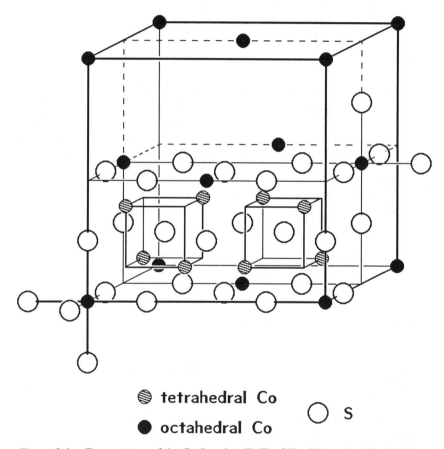

⊜ **tetrahedral Co**

● **octahedral Co**   ◯ **S**

Figure 9.1. Two ocatants of the Co_9S_8 unit cell. The full cell contains 12 cobalt ions on tetrahedral sites. The octahedral sites at the eight corners and six face centers contribute a total of four octahedral cobalt ions per unit cell, since corner positions are shared among eight unit cells and face positions are shared between two.

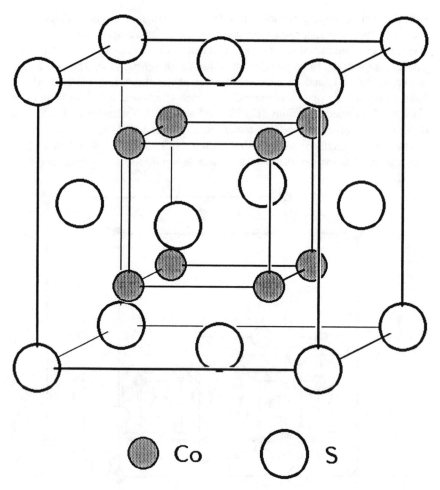

Figure 9.2. The cubic cluster in Co_9S_8.

$a = 5.738(2)$ Å and $c = 7.126(2)$ Å (15). There are three nickel atoms and two sulfur atoms per unit cell as shown in Fig. 9.3. The nickel atoms are in pseudotetrahedral positions formed by their coordination with four sulfur atoms.

Fe_7S_8 can be considered a derivative of FeS which has the layered NiAs structure (Fig 9.4). In Fe_7S_8 one-eighth of the iron atoms have been removed. Random distribution of the vacancies results in a unit cell that is hexagonal. However, when the vacancies order among alternate layers of iron atoms, the unit cell is monoclinic (Fig. 9.5). Slow cooling or annealing of Fe_7S_8 samples results in ordering of these vacancies.

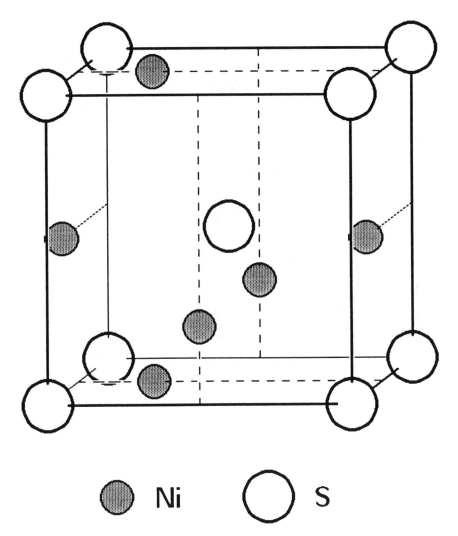

Figure 9.3. The unit cell of Ni_3S_2.

b. Low-Temperature Preparation of Sulfides with the Pyrite Structure

Bouchard (17) has used coprecipitated mixed crystal sulfates to prepare solid solutions of the type $Fe_xCo_{1-x}S_2$, $Co_xNi_{1-x}S_2$ and $Cu_xNi_{1-x}S_2$. The mixed sulfates were precipitated out of aqueous solution by their addition to 10 times their volume of acetone. The products were filtered and heated at 300°C for 6 hours in an $H_2S/N_2 = 1/1$ atmosphere. The resulting sulfides were quenched to room

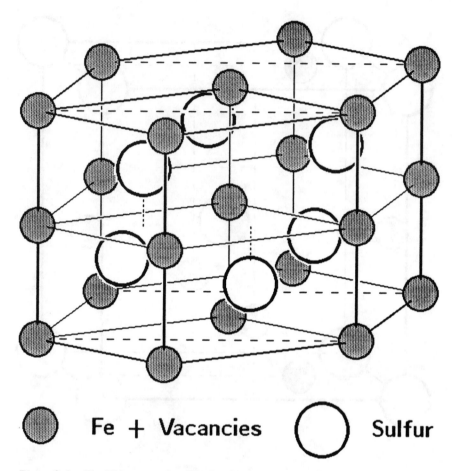

Fe + Vacancies   **Sulfur**

Figure 9.4. The NiAs structure of Fe_7S_8 with random vacancies.

temperature. The particle size was determined by X-ray diffraction line broadening to be ≈500 Å. The critical step of the preparation is the low-temperature treatment with H_2S. Such products gave single-phase well-crystallized sulfides when heated to higher temperatures.

Chianelli and Dines (18) reported a novel technique for the preparation of a number of transition metal dichalcogenides. These sulfides of group IV, V, and VIB were synthesized in nonaqueous solutions at room temperature by reaction between the anhydrous transition metal chloride and either lithium sulfide or ammonium hydrogen sulfide. The products obtained by this method were poorly crystallized and had high surface areas. The synthetic method can be represented by the following equations:

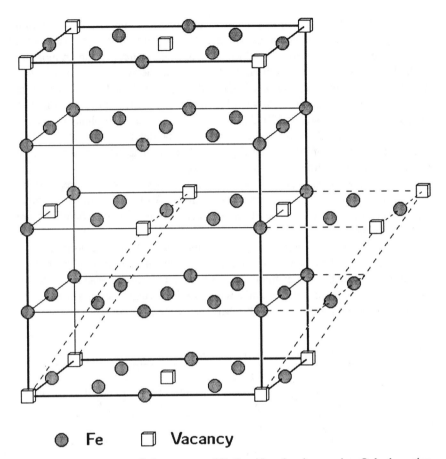

● **Fe**   ☐ **Vacancy**

Figure 9.5. The monoclinic structure of Fe_7S_8 with ordered vacancies. Only the cation sites are shown.

$$MX_4 + 2A_2S \xrightarrow[\text{ambient temp and pressure}]{\text{solvent}} MS_2 + 4AX \qquad (1)$$

and

$$MX_5 + (5/2)A_2S \xrightarrow[\text{ambient temp and pressure}]{\text{solvent}} MS_2 + 5AX + (1/2)S \qquad (2)$$

In the first reaction, the metal(IV) halide reacts with a sulfide source to form a metal disulfide and a halide salt. The disulfides ZrS_2, HfS_2, VS_2, MoS_2 and TiS_2 have been prepared by this method. In the second redox reaction, the metal(V) state is reduced to metal(IV), and examples of this process are TaS_2 and

NbS_2. A polar solvent free of oxygen and water is used to prevent the formation of a transition metal oxide or hydroxide. The solvents that were used were ethylacetate or tetrahydrofuran, and the chalcogenide source was an ammonium or alkali metal sulfide. Passaretti et al. (19) reported the synthesis of amorphous RuS_2 by the reaction of anhydrous $RuCl_3$ and NH_4HS in 2-methoxyethyl ether at room temperature. The product was filtered and extracted with methanol, then heated in a stream of H_2S at 250°C. The final product was amorphous but began to crystallize as a pyrite phase when annealed at temperatures greater than 350°C. The extent of crystallinity was directly related to the annealing temperature. At 825°C, a well-crystallized product was obtained.

Amorphous RuS_2 as well as OsS_2, PtS_2, and PdS_2 were also prepared by the reaction of the anhydrous hexachlorometallate(IV) with hydrogen sulfide at various temperatures (20). The starting materials and reaction conditions are given in Table 9.1.

A sample of CoS_2 was prepared by sulfurizing $Co(NH_3)_6Cl_2$ for 2 hours at room temperature (21). The ammonium chloride was removed by extracting the product in a soxhlet extractor with methanol for 24 hours. The product was annealed for 4 days at 800°C in a sealed silica tube with 10% by weight of excess sulfur. The sulfur to cobalt ratio was obtained by thermogravimetric analysis to give a value of 1.99:1.

Table 9.1. Preparation of Platinum Metal Dichalcogenides

Disulfide	Starting material	Temp of reaction (°C)	Reaction time (hours)	Temp of anneal (°C)	Length anneal (days)
RuS_2	$(NH_4)_2RuCl_6$	180	4	800	4
OsS_2	$(NH_4)_2OsCl_6$	220	3	800	4
PtS_2	$(NH_4)_2PdCl_6$	110	2	750	6
PdS_2	$(NH_4)_2PtCl_6$	130	1	700	5

Schleich prepared a number of amorphous transition metal sulfides by a novel low temperature technique (22–24). The process involves the reaction between metal halides MX_n (M = Ta, Nb, Mo; X = Cl, F) and organic sulfur compounds such as hexamethyldisilthane (HMDST), di-*tert*-butyldisulfide (DTBDS), di-*tert*-butylsulfide (DTBS), and *tert*-butylmercaptan (TBMC). Crystalline sulfides were obtained by heating the amorphous powders in vacuum or by direct reaction of the amorphous products with sulfur. These reactions were carried out in evacuated silica tubes. Different stoichiometries were obtained by changing the temperature and partial pressure of sulfur.

Kaner and co-workers (25) reported a solid-state metathesis reaction between molybdenum(V) chloride and sodium sulfide to form MoS_2:

$$MoCl_5 + (5/2)Na_2S \rightarrow MoS_2 + 5NaCl + (1/2)S$$

This technique yields a well-crystallized product of high quality in a matter of seconds. This method is also applicable to the synthesis of other dichalcogenides, e.g., WS_2, WSe_2, and $MoSe_2$.

It has recently been discovered (26) that metal–dithiocarbamate complexes may be used as precursors to metal sulfide materials as shown in the following equation:

$$\begin{matrix} M(Et_2dtc)_2 \\ \text{or} \\ M(Et_2dtc)_3 \end{matrix} \xrightarrow{\ H_2/H_2S\ } MS_x + \text{organic decomposition products}$$

The dithiocarbamates are well-known coordination complexes (27) and are easily prepared. They decompose under hydrogen sulfide at temperatures slightly below their melting points. Further heating in an atmosphere of H_2/H_2S in the range of 400–600°C produces metal sulfide phases that correspond to those obtained by direct combination of the elements at higher temperatures. Among the sulfides that can be prepared by this method are ZnS, Ni_3S_2, and Co_9S_8.

B. 3*d* Transition Metal Disulfides with the Pyrite Structure

The 3*d* transition metals form disulfides that crystallize with the pyrite structure. These compounds have received considerable attention because their conductivity ranges from semiconductor to metallic behavior and they evidence a wide range of magnetic properties. One of the chalcogenides, namely FeS_2, can also adopt the marcasite structure. The pyrite structure is shown in Fig. 9.6 and contains discrete S_2 groups. There is a close resemblance of the pyrite structure with that of sodium chloride; the transition metal atoms and the S_2 centers occupy the Na^+ and Cl^- positions. The iron atoms are on octahedral sites with six nearest neighbor sulfur atoms, whereas the sulfur coordination is tetrahedral consisting of one sulfur and three iron atoms.

Bouchard (28) has shown that single crystals of the MS_2 (M = Fe, Co, Ni) pyrites can be grown by chemical vapor transport using chlorine as the transport agent. However, Butler and Bouchard (29) concluded that chlorine cannot be used as a transport agent for the crystal growth of $Fe_xCo_{1-x}S_2$, and that bromine is the best transport agent for solid solutions between any pair of the members FeS_2, CoS_2, and NiS_2. Both thermodynamic predictions as well as experimental observations have shown that bromine is the best transport agent. Their observed rate growths were in agreement with their calculated values.

A most interesting synthetic procedure was used by Bouchard (17) for the preparation of polycrystalline samples of pyrite solid solutions. The method was suggested by the work of Delafosse et al. (5) who synthesized CoS_2 and NiS_2 by heating the corresponding sulfates in H_2S at 450° and 227°C, respectively. This is a particularly efficient way of preparing the solid solutions since the hydrated

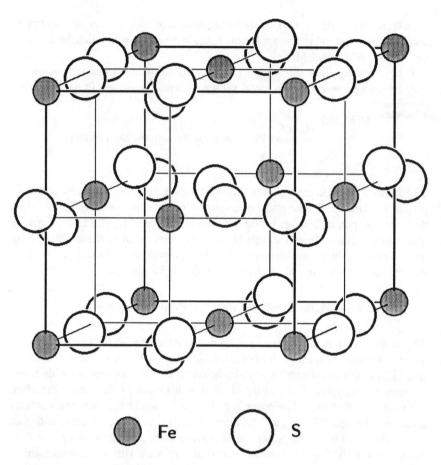

Fe   S

Figure 9.6. The pyrite structure.

sulfates of Fe, Co, Ni, and Cu are isomorphous and therefore can form mixed crystals. If pure products are desired, the transition metals of high purity can be weighed out and dissolved in dilute sulfuric acid to form the mixed sulfates.

It has been shown from both magnetic and electrical measurements that the transition metals Fe, Co, Ni, and Cu are present as low spin divalent ions in the MS_2 pyrites. FeS_2, NiS_2, and ZnS_2 behave as semiconductors (0, 2, and 4 e_g electrons), whereas CoS_2 and CuS_2 (1, 3 e_g electrons) are metallic. It was proposed by Jarrett et al. (30) and Bither et al. (31) that the two e_g levels of the octahedrally coordinated metal species broaden into a band. A one-electron energy diagram for a $3d$ cation surrounded by six sulfur atoms is given in Fig. 9.7. Such diagrams have been utilized by many investigators (29–32).

FeS_2 is a diamagnetic semiconductor. Hence, the antibonding e_g levels are shifted upward (Fig. 9.7) far enough so that all six $3d$ electrons occupy the t_{2g}

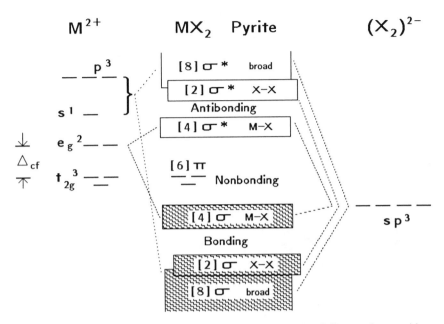

Figure 9.7. Schematic one-electron, one-molecule energy level diagram for transition metal pyrites.

levels. It was observed that CoS_2 $(3d^7)$ is ferromagnetic. $T_c = 115$ K and the saturation moment corresponds to 0.8 unpaired electrons per cobalt. The Co(II) ion is, therefore, $(t_{2g})^6 (e_g)^1$ and is in the low spin state. The metallic conductivity observed for CoS_2 indicates that the e_g electrons are nonlocalized. For NiS_2 (Fig. 9.8), the intraatomic exchange is larger than the band width so that the energy levels are split into spin-up and spin-down states. Hence, NiS_2 behaves as a semiconductor with an activation energy corresponding to that required for the transfer of an electron from the filled spin-up band to the empty spin-down band. The measured magnetic and electrical properties of the $3d$ transition metal sulfides are consistent with the simple concept of the band structure of these compounds described by Goodenough (32).

The ferromagnetic pyrite CoS_2 was further studied by Johnson and Wold (3). In their studies, selenium was substituted for sulfur. The system $CoS_{2-x}Se_x$ has the advantage that the magnetic atom is not changed and, for all compositions, the number of electrons in the $\sigma^*(e_g)$ band should be the same on the basis of any reasonable bonding model. Furthermore, it might be expected that substitution of the more polarizable Se atom for S should lead to a greater degree of covalence, hence, increased band width for the $\sigma^*(e_g)$ band. It was shown by Johnson and Wold (3) as well as Adachi et al. (33) that selenium substitution introduces strong antiferromagnetic interactions between cobalt atoms. First-order ferromagnetic

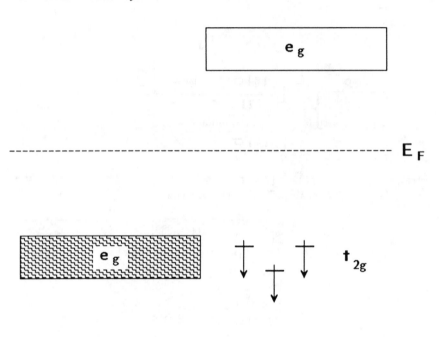

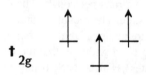

Figure 9.8. Schematic one-electron, one-molecule energy level diagram for NiS_2.

to paramagnetic transitions for some of the solid solutions were suggested by both the temperature and field dependence of the magnetization in the neighborhood of T_c. Furthermore, compositions between $x = 0.20$ (ferromagnetic) and $x = 0.30$ (antiferromagnetic) appear to exhibit noncolinear structures.

A larger influence of the anion on the type of magnetic order that prevails is exhibited by this system. The exact role of the anion is not clear, however, since the observed properties cannot unambiguously separate the effects of volume changes, slight configurational changes, and increased covalency (increased band width) as Se is substituted for S. However, the electronic configuration of the cobalt atom is not significantly changed across the series. What does change is the magnitude, and/or signs, of the magnetic interactions.

C. Layer Structures

The layered dichalcogenides can be divided into two classes depending on whether the metal atom is on an octahedral or trigonal prismatic site. Octahedral coordination is associated with the CdI_2 structure and the sulfides TiS_2, ZrS_2, and PtS_2 adopt this C6 structure. The structure consists of planar (MS_2) units with MS_6 octahedra joined at the edges and the sheets are held together by van der Waals bonding (Fig. 9.9). Jeannin and Benard (34,35) have indicated that the range of nonstoichiometry for TiS_2 is $Ti_{1.04}S_2$–$Ti_{1.105}S_2$. For the nonstoichiometric compounds, the lattice parameters of the hexagonal unit cell increase with increased deviation from TiS_2. Single crystals of TiS_2 and ZrS_2 have been grown by chemical vapor transport (36) using iodine as the transport agent. They reported that both TiS_2 and ZrS_2 were nearly metallic degenerate semiconductors at room temperature. Thompson et al. (37) investigated the electrical and optical properties of TiS_2 and indicated that the observed metallic behavior of TiS_2 can be attributed to the overlap of the d band with the valence band. This is in agreement with the results of Takeuchi and Katsuta (38) who have suggested that there is a small degree of overlap of the bottom of the d band with the top of the valence band. However, in ZrS_2 (and HfS_2), the d band is at a higher energy level and no overlap occurs. Hence, ZrS_2 and HfS_2 do not show metallic behavior. PtS_2 has been characterized as a diamagnetic semiconductor (39–41). Finley et al. (42) have reported the growth of well-formed crystals of $Pt_{0.97}S_2$ by chemical vapor transport. A combination of phosphorus and chlorine was used as the

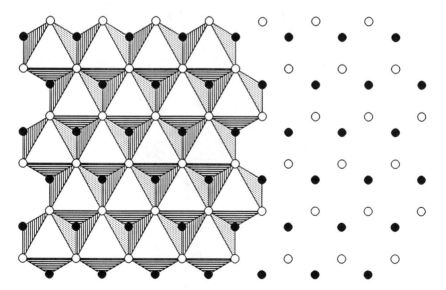

Figure 9.9. The CdI_2 structure of TiS_2 and related dichalcogenides.

transport agent. The crystals were found to be diamagnetic with a susceptibility of $-31(2) \times 10^{-6}$ emu/mol at 77 K and showed semiconducting behavior with a band gap of 0.20(2) eV. The composition of the crystals was ascertained from precise density determinations and refined cell constants.

MoS_2, as well as the other dichalcogenides of Mo and W, crystallizes as a layer-type structure (Figs. 9.10 and 9.11) (43–45). The basic structural units of MoS_2 consist of S–Mo–S sandwiches in which the metal is in a trigonal prismatic coordination of sulfur atoms (46). The sandwich layers have hexagonal symmetry and are stacked along the crystallographic c-direction. Differences in stacking of

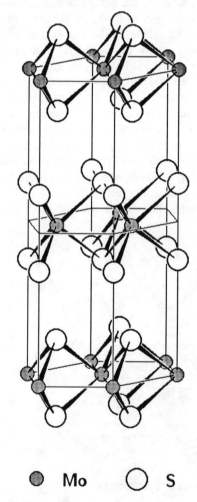

● **Mo**   ○ **S**

Figure 9.10. The hexagonal unit cell of the 2H polytype of MoS_2.

(a)   (b)

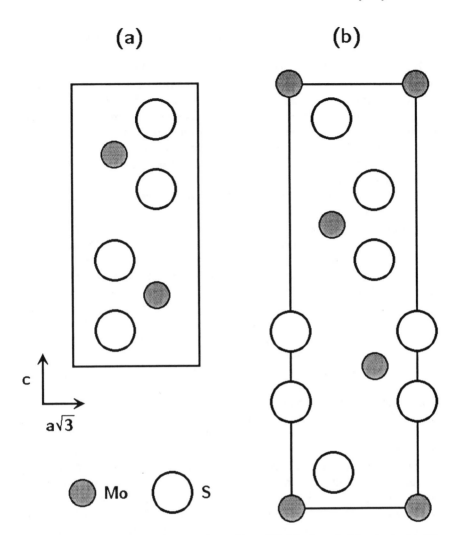

Figure 9.11. Comparison of the *c* axis stacking of S–Mo–S sandwiches in the (a) 2H and (b) 3R polytypes of MoS_2.

these layers leads to different polytypes. For MoS_2, the 2H and 3R polytypes are the most common. The 2H polytype consists of two S–Mo–S sandwiches within a hexagonal cell (Figs. 9.10 and 9.11a). For the 3R polytype, three MoS_2 layers are found in a unit, all having rhombohedral symmetry (Fig. 9.11b).

Regardless of the stacking sequence, considerable anisotropy exists in the chemical bonding (45). Weak van der Waals forces bind the individual S–Mo–S sandwiches to one another, producing an S–S bond distance of 3.49 Å. The distance between MoS_2 layers is 2.96 Å and is commonly referred to as the van der Waals gap. Within a S–Mo–S sheet the bonding is predominantly covalent with Mo–S and Mo–Mo distances of 2.41 and 3.16 Å, respectively.

The anisotropic behavior observed in the properties of MoS_2 can be related to the anisotropy found in the chemical bonding (45). The weak van der Waals' forces between MoS_2 layers allow for easy cleavage. Lubricating properties comparable to those of graphite are observed for MoS_2 (47) presumably because the MoS_2 layers can slide against each other in directions normal to the crystallographic c axis. The electrical conduction within covalently bonded S–Mo–S layers (perpendicular to c) is considerably higher than that observed between these layers (parallel to c).

MoS_2 crystals can be grown by chemical vapor transport using bromine or iodine as the transport agent (48). The physical properties of MoS_2 are summarized in two excellent review papers by Wilson and Yoffe (45) and Opalovskii and Fedorov (49). MoS_2 is a diamagnetic semiconductor. In MoS_2, the degeneracy of the molybdenum $4d$-orbitals is partially removed by the trigonal prismatic coordination (D_{3h} symmetry) of sulfur atoms about molybdenum (Fig. 9.12a) (50). Ligand field splitting of the d-orbitals (Fig. 9.12b) produces a lowest lying $4d$ level having A_1' symmetry. The degenerate $4d_{xy}$ and $4d_{x^2-y^2}$ orbitals are of next highest energy and have E' symmetry. The degenerate pair of $4d_{xz}$ and $4d_{yz}$ orbitals have E'' symmetry and are of highest energy. For Mo(IV), there is a $3d^2$ electron configuration, resulting in a filled A_1' level.

For semiconducting MoS_2, d-orbitals of similar symmetry from neighboring molybdenum atoms can interact to form energy bands (45,50). An energy band diagram of MoS_2 can be compared with that of the isostructural compound NbS_2 (Fig. 9.13) (51). The valence band, V, of both dichalcogenides is broad and made up primarily of sulfur s- and p-orbitals. The conduction band, C, is also broad and consists primarily of metal s- and p-orbitals. The d bands, however, are narrow and have energies that overlap the energies of the valence and conduction bands. The positions of the d bands with respect to the valence and conduction bands have been the subject of several investigations (52-58). The A_1' band is of lowest energy (52) and lies close to the valence band. The E' and E'' bands are of higher energy but only the E'' overlaps the conduction band. Spin orbit coupling causes splitting of both the E' and E'' bands (50). The number of electrons in MoS_2 is sufficient to fill completely the A_1' band. However, NbS_2 has one less electron per metal atom so that the A_1' band is only half-filled. Consequently, NbS_2 shows metallic conduction.

The layer-type structure of MoS_2 and NbS_2 leads to the possible occurrence of intercalation. This involves the insertion of foreign atoms, ions, or molecules into the van der Waals gap with the expansion of the host structure in a direction normal to the layers (59,60). These foreign species can occupy the vacant octahedral (or tetrahedral) sites available within the van der Waals gap. There has been no substantiated report of intercalation into MoS_2. Haering et al. (61) reported the existence of Li_xMoS_2 but the nature of this phase has not been completely

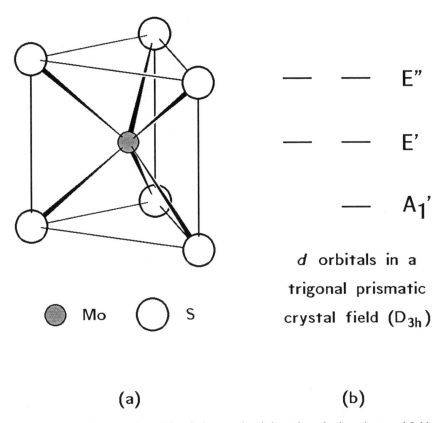

— — E"

— — E'

— A₁'

d orbitals in a
trigonal prismatic
crystal field (D_{3h})

● Mo ○ S

(a)   (b)

Figure 9.12.   Mo-S bonds and d-orbital energy levels in a trigonal prismatic crystal field.

understood. The sorption of hydrogen by MoS_2 to give H_xMoS_2 occurred only when the MoS_2 was freshly prepared as a high surface area catalyst. Single crystals of MoS_2 showed a very small uptake of hydrogen (62). NbS_2 does, however, form intercalation compounds (36,60–68).

Differences observed in the intercalation behavior of MoS_2 and NbS_2 can be related to differences in the band structure of these two dichalcogenides (Fig. 9.13) (50,51). Intercalation of $3d$ metals, for example, involves electron donation to an empty or partially filled band of the host material. For MoS_2, electron donation would require filling the empty E' band which is not favorable from an energy standpoint. However, for NbS_2, electron donation from $3d$ metals can readily occur by the filling of the A_1' partially occupied band.

Although tin is not a transition element, the chemistry of the layer compound SnS_2 is interesting and deserves mention. SnS_2 is a semiconductor which crystallizes in the layered hexagonal CdI_2 structure, space group $C6$. The structural unit can be described as two layers of hexagonal, close-packed sulfur anions with a

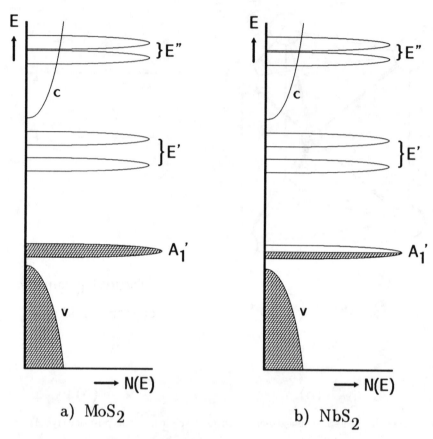

Figure 9.13. Comparison of the band structures of (a) MoS_2 and (b) NbS_2 (51).

layer of tin cations sandwiched between them. The tin cations are octahedrally coordinated by six sulfur anions. Adjacent sulfur layers are bound by weak van der Waals interactions.

There have been numerous studies (36,68–75) of SnS_2 crystals prepared by chemical vapor transport, in which iodine was the predominant transport agent used. These included structural investigations (72–75) of polytypes formed from different stacking arrangements of the tin and sulfur layers, as well as electrical studies. The electrical resistivities of crystals from these reports varied several orders of magnitude from 10^2 to 10^{12} Ω-cm, and the activation energies ranged from 0.1 to 0.05 eV.

Single crystals were grown by Kourtakis et al. (76) who utilized both vapor transport and chemical vapor transport using chlorine as the transport agent. Chemical vapor transport was the procedure predominantly used in previous reports on the preparation of SnS_2 crystals. There are large discrepancies in the electrical properties of crystals in those studies. Two important factors which may cause these discrepancies are (a) the inclusion of halogen impurity in the SnS_2 crystal and (b) slight departures from stoichiometry, i.e., a metal to sulfur ratio greater than 1:2. The comparative study by Kourtakis et al. (76) examined the impact of these two factors on the properties of SnS_2.

Crystals prepared by both vapor transport and chemical vapor transport were indexed on a simple hexagonal unit cell with $a = 3.649(2)$ Å, $c = 5.902(2)$ Å or $5.900(2)$ Å as shown in Table 9.2. These values were in close agreement with those of Mikkelsen (77) [$a = 3.649$ Å and $c = 5.884(5)$ Å] and Whitehouse and Balchin (72) [$a = 3.643(2)$ Å and $c = 5.894(5)$ Å]. However, they differed from those reported by Greenway and Nitsche (68) [$a = 3.639(5)$ Å and $c = 5.884(5)$ Å] and Conroy and Park (36) [$a = 3.644(4)$ Å and c $= 5.884(4)$ Å].

The electrical properties of crystals grown by vapor transport under different conditions were investigated, since crystals grown by this method are free of halogen impurities. Therefore, any anomalies in the electrical properties can be directly attributed to slight deviations from stoichiometry. SnS_2 is a low resistivity n-type crystal when grown with charge-growth temperatures of 750–700°C, as indicated in Table 9.3. In these reactions, stoichiometric amounts of sulfur and tin were used as the charge. The material exhibits a resistivity $\rho = 4.5(5)$ Ω-cm at 25°C. Crystals can be prepared in which the majority of carriers are p-type when grown at 650–600°C with 5% excess sulfur added to the charge. The p-type crystals showed high resistivity $(\rho > 10^7$ Ω-cm) at 25°C. The high resistivity of p-type SnS_2 indicates that this material is stoichiometric, whereas the lower resistivity of n-type SnS_2 samples shows that the tin to sulfur ratio deviates from 1:2. As shown in Table 9.4, the n-type crystals were annealed at 500 and 600°C in a sulfur atmosphere to produce a more stoichiometric material. The resistivity was shown, in fact, to increase from 4.5 to 2 x 10^5 Ω-cm, demonstrating that nonstoichiometry is predominantly responsible for the low resistivity of SnS_2 prepared at higher temperatures. This shows that both the temperature used in

Table 9.2. X-Ray and Preparative Data for SnS_2

Process	Gradient (°C)	Cell Parameters	
		a(Å)	c(Å)
CVT (chlorine)	730–680	3.649(2)	5.902(2)
CVT (chlorine)	640–590	3.649(2)	5.900(2)
Vapor transport (5% excess sulfur)	650–600	3.652(2)	5.902(2)
Vapor transport	750–700	3.651(2)	5.904(2)
CVT (Ref. 68)	800–700	3.639(5)	5.884(5)
CVT (Ref. 36)	700–600	3.644(4)	5.884(4)
CVT (Ref. 69)	690–650	Not reported	Not reported

Table 9.3. Electrical and Optical Properties of SnS_2 Crystals

Process	Gradient (°C)	$\rho(25°C)$ (Ω-cm)	$E_a(eV)$	Carrier type	$E_g(eV)$
Vapor transport	750–700	4.5	—	n	2.28(5)
Vapor transport	640–590	10^4	—	n	—
Vapor transport (5% excess sulfur)	650–600	$>10^7$	—	p	2.28(5)
CVT,Cl_2	730–680	3	0.06(1)	n	2.22(5)
	640–590	5	0.06(1)	n	2.22(5)
CVT (Ref.68)	800–700	—	—	—	2.21
CVT (Ref.36)	700–600	10^9	—	n	2.20
CVT (Ref.69)	690–650	10^2–10^{12}	0.05–0.4	n	—

the growth and the stoichiometry of the charge affect the degree of nonstoichiometry in SnS_2.

The electrical properties of crystals grown by chemical vapor transport were also measured to determine the effect of the halogen transport agent, Cl_2 (Table 9.3). Crystals grown by CVT are n-type semiconductors with $\rho = 5\ \Omega$-cm at 25°C and are not appreciably affected by the growth temperature. Neutron activation analysis shows 22(6) ppm chlorine to be present in crystals grown where the charge-growth temperatures are 640–590°C. Crystals grown by this method, when annealed under the same conditions as those used for vapor transport crystals, show no significant change in resistivity (Table 9.4). This indicates that

Table 9.4. Electrical Properties of Annealed SnS_2 Vapor Grown Crystals

Process	Anneal	Resistivity ρ (25°C) (Ω-cm)	Mobility $\mu(25°C)$ (cm²/volt sec)	Carrier type
Vapor transport (750–700°C)	As grown	4.5(5)	26(5)	n
	500°C (<1 atm sulfur)	6.9(5)	46(5)	n
	500°C (2 atm sulfur)	8×10^4	—	n
	600°C (3 atm sulfur)	2×10^5	—	n
CVT,Cl_2 (640–590°C)	As grown	5.0	21(1)	n
	500°C (<1 atm sulfur)	4.5(5)	19(1)	n
	500°C (2 atm sulfur)	6.9(5)	9.0(1)	n

nonstoichiometry does not contribute appreciably to the magnitude of electrical resistivity. Instead, the presence of halogen impurity must account for the low resistivity of this material.

Further differences between the electrical properties of crystals grown by CVT and vapor transport are shown by the temperature dependence of the resistivity (Fig. 9.14). Crystals grown by vapor transport, 750–700°C (Fig. 9.14), show resistivities that decrease with decreasing temperature, indicative of a degenerate semiconductor. Such resistivities are due to scattering rather than carrier activation. The donor states introduced by slight deviations in stoichiometry lie close to the conduction band edge and are ionized. In contrast, crystals grown by chemical vapor transport show classical semiconductive behavior, i.e., the resistivity increases with decreasing temperature (Fig. 9.14). It has already been noted that these materials are not sulfur deficient, since their resistivities are unaffected by annealing experiments in sulfur. This is further borne out by the graph of log ρ vs $1/T$ (Fig. 9.14), which shows no deviation from the ideally linear case of a classic semiconductor. If donor states arising from nonstoichiometry were also present in this material, then the linearity of this plot would be affected. The

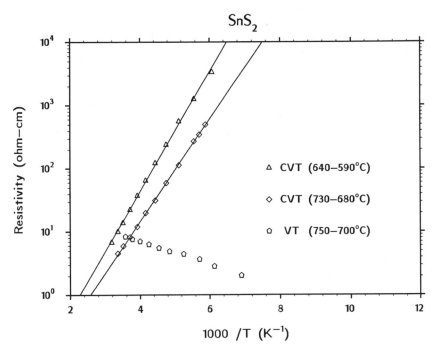

Figure 9.14. Resistivity as a function of temperature for SnS$_2$ crystals grown by vapor transport (VT) and by chemical vapor transport (CVT) with various charge-growth gradients.

activation energy obtained from these plots for crystals grown by CVT is 0.06(1) eV. This suggests that the chlorine impurity is solely responsible for the donor states, and these states lie slightly below the conduction band. These results, therefore, explain the discrepancies in the electrical properties that have been reported in the literature.

D. Intercalation Compounds

In solid state chemistry, intercalation compounds are formed when foreign ions, atoms, or molecules are inserted between the layers of a crystalline compound. The original structure of the host compound is stable in the absence of the foreign species. The intercalation compounds formed by inserting the alkali metals into the disulfides TiS_2, NbS_2, and TaS_2 have been well studied in recent years. In 1969, Weiss and Ruthhardt (78) reported the formation of intercalation complexes between TiS_2 and both aqueous hydrazine and alykotic amides. Gamble et al. (67) showed that intercalation into the layered disulfides and diselenides of Group IVB and VB transition metal and molecular Lewis bases is quite general. These compounds show interesting electronic properties, e.g., the niobium and tantalum dichalcogenide compounds are superconductors. The interaction between the intercalate and the chalcogenide can be modified by changing the molecular base that is used to form the complex. These changes also affect the observed electronic properties.

The NH_3 intercalation compounds of TiS_2, ZrS_2, NbS_2, and TaS_2 can be prepared using gaseous or liquid ammonia (67). The compounds formed with aqueous NH_3 are hydrates with the stoichiometry $(NH_3)_{1/3}(H_2O)_{2/3}TaS_2$ (67,79,80). It is also possible to intercalate NH_3 by the electrolysis of an ammonium salt at a dichalcogenide electrode (81,82). In such interchelates, the volume of the dichalcogenide is increased by 50% and the interlayer distance increases by 3 Å.

Whittingham (83) reviewed the use of Li_xTiS_2 as a potential cathode material for high density batteries. It was shown that there is essentially no structural change except for an expansion perpendicular to the basal planes. In the Li/TiS_2 system, only a single phase exists from the completely charged to uncharged states. The intercalation of lithium into TiS_2 may be reversed many times even at room temperature. The reaction can be represented by

$$x\text{Li} + \text{TiS}_2 \rightleftharpoons \text{Li}_x\text{TiS}_2 \text{ for } 0 \le x \le 1$$

Whittingham (84) showed that both the lattice expansion and the cell emf change in a continuous manner with the state of discharge. In the system Li_xTiS_2, the lithium species enters the van der Waals gap and no bonds within the TiS_2 layers are perturbed.

The insertion of an alkali metal into a transition metal dichalcogenide involves the acceptance of electron(s) into the d band (see Fig. 9.13). Simultaneously, the alkali metal diffuses into unoccupied sites between the MX_2 slabs. Obviously, an empty or nearly empty d-conduction band can receive up to six electrons from each octahedrally coordinated metal atom. Hence, such an intercalation compound will contain the largest amount of alkali metal compound. If the metal atom occupies a trigonal prismatic site, then the A_2' band (see Fig. 9.13) can hold only two electrons per metal atom.

E. Chevrel Phases

In 1971, Chevrel et al. (85) reported the preparation and structure of a number of ternary molybdenum compounds with unusual electronic properties. In the compounds with the stoichiometry $M_xMo_6X_8$ (M = Li, Mn, Fe, Cd, Pb; X = S, Se, Te) there are formed Mo_6 octahedral clusters (86). These compounds with $4 \geqq x \geqq 0$ are of great interest because many are superconducting. The basic building blocks are Mo_6X_8 clusters. The eight X atoms are at the corners of a cube with each of the six Mo atoms near a face center of the cube. This is shown in Fig. 9.15. The homogeneity range x of the M atoms depends on their size. For atoms which are relatively small, x can have values up to 4. Such atoms, e.g., Cu, are found among two sets of six tetrahedral sites off the $\bar{3}$ axis (87). Large M atoms (Pb, Sn) are located on the $\bar{3}$ axis and the value of x is one (Fig. 9.15). For Chevrel phases, where M = light rare earth, the composition approaches $RE_{1.0}Mo_6S_8$, but for the heavy earths, the stoichiometry is closer to $RE_{1.2}Mo_6S_8$ (88).

For the Mo_6S_8 cluster, the six molybdenum atoms have a total of 36 available valence electrons, i.e., six electrons for each molybdenum atom. The formation of eight sulfide (S^{2-}) ions requires 16 electrons and the remaining 20 electrons completely fill the bonding d-orbitals. There is also a nonbonding e_g level that can accommodate a total of four additional electrons per cluster. These electrons can be provided by the intercalation of a metal such as lithium. The formation of a nonstoichiometric phase $Li_xMo_6S_8$, where x has an upper limit of 3.6, agrees with the model that limits the nonbonding e_g level to four electrons (24 free electrons per Mo_6S_8 cluster). In a solid with long-range order, Mo_6S_8 clusters are extended so that band formation occurs. However, the above electron counting scheme is still valid.

Many of these compounds can be prepared by reacting the elements in evacuated silica tubes. Manipulation of the air-sensitive sulfides, e.g., the alkaline and rare earth sulfides, must be carried out in a deoxygenated, dry-argon filled glove box (85). Lithium insertion can be accomplished by reaction with n-butyl lithium at ambient temperature followed by annealing at 450°C for five days to achieve uniform distribution of the lithium (89).

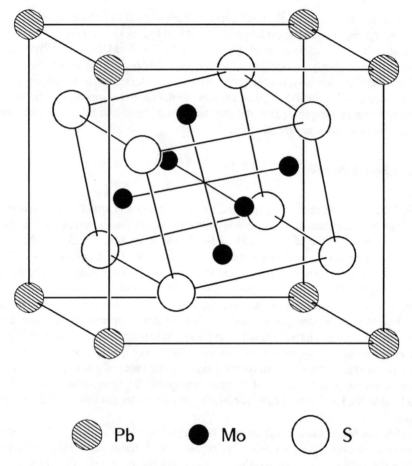

Pb ● Mo ○ S

Figure 9.15. Structure of a Chevrel compound showing the canted Mo_6S_8 unit in a slightly distorted cube of Pb atoms.

The lithiated compositions show a large change in the lattice parameters as x goes from 0 to 3 in $Li_xMo_6S_8$. The phases $LiMo_6S_8$ and $LiMo_6Se_8$ superconduct at 5 and 3.94 K, but the lithium-rich phase ($x = 3.2$) does not show superconducting behavior. This is not surprising since, when the total number of free electrons for each Mo_6S_8 cluster reaches 24, the phase is expected to show semiconducting properties (90,91). Unfortunately, the mobility of lithium within the Mo_6S_8 matrix is high at 300 K (92). The lithium can diffuse out of the channels by reacting with water vapor in the air to form LiOH.

References

1. H. Nahigian, J. Steger, R. J. Arnott, and A. Wold, *J. Phys. Chem. Solids*, **35**, 1349 (1974).

2. J. C. Mikkelson and A. Wold, *J. Solid State Chem.*, **3**, 39 (1971).

3. V. Johnson and A. Wold, *J. Solid State Chem.*, **2**, 209 (1970).

4. A. H. Thompson, F. R. Gamble, and C. R. Symon, *Mat. Res. Bull.*, **10**, 915 (1975).

5. D. Delafosse and P. Barret, *C. R. Acad. Sci. Paris*, **251**, 2964 (1960).

6. D. Delafosse, M. Abon, and P. Barrett, *Bull. Soc. Chim. France*, 1110-11 (1961).

7. D. Delafosse and P. Barret, *C. R. Acad. Sci. Paris*, **252**, 280 (1961).

8. D. Delafosse and P. Barret, *C. R. Acad. Sci. Paris*, **252**, 888 (1961).

9. K. Kim, K. Dwight, A. Wold, and R. R. Chianelli, *Mat. Res. Bull.*, **16**, 1319 (1981).

10. G. Kullerud and R. A. Yund, *J. Petrol.*, **3**(1), 126 (1962).

11. F. K. Lotgering, *Philips Res. Rep.*, **11**, 190 (1956).

12. L. Néel, *Rev. Mod. Phys.*, **25**, 58 (1953).

13. E. F. Bertaut, *Bull. Soc. Fr. Mineral. Crystallogr.*, **79**, 276 (1956).

14. R. Benoit, and C. R. Hebd, *Seanc. Acad. Sci.*, **234**, 2174 (1952).

15. D. M. Pasquariello, R. Kershaw, J. D. Passaretti, K. Dwight, and A. Wold, *Inorg. Chem.*, **23**, 872 (1984).

16. V. Rajamani and C. T. Prewitt, *Can. Mineral.*, **13**, 75 (1975).

17. R. J. Bouchard, *Mat. Res. Bull.*, **3**, 563 (1968).

18. R. R. Chianelli and M. B. Dines, *J. Inorg. Chem.*, **17**, 2758 (1978).

19. J. D. Passaretti, R. C. Collins and A. Wold, *Mat. Res. Bull.*, **14**, 1167 (1979).

20. J. D. Passaretti, R. B. Kaner, R. Kershaw, and A. Wold, *Inorg. Chem.*, **20**, 501 (1981).

21. J. D. Passaretti, K. Dwight, A. Wold, W. J. Croft, and R. R. Chianelli, *Inorg. Chem.*, **20**, 2631 (1981).

22. D. M. Schleich and M. J. Martin, *J. Solid State Chem.*, **64**, 359 (1986).

23. A. Bensalem and D. M. Schleich, *Mat. Res. Bull.*, **23**, 857 (1988).

24. A. Bensalem and D. M. Schleich, *Mat. Res. Bull.*, **25**, 349 (1990).

25. P. R. Bonneau, R. F. Jarvis, Jr., and R. B. Kaner, *Nature (London)*, **349** 510, (1991).

26. R. D. Pike, R. Kershaw, K. Dwight, A. Wold, T. Blanton, A. Wernberg, and H. Gysling. To be published in *Thin Solid Films*, Feb. 1993.

27. D. Coucouvanis, *Prog. Inorg. Chem.*, **26**, 301 (1979).

28. R. Bouchard, *J. Cryst. Growth*, **2**, 40 (1968).

29. S. R. Butler and R. J. Bouchard, *J. Cryst. Growth*, **10**, 163 (1971).

30. H. S. Jarrett, W. H. Cloud, R. J. Bouchard, S. R. Butler, C. G. Fredericks, and J. L. Gillson, *Phys. Rev. Lett.*, **21**(9), 617 (1968).

31. T. A. Bither, R. J. Bouchard, W. H. Cloud, P. C. Donohue, and W. S. Siemons, *Inorg. Chem.*, **7**(11), 2208 (1968).

32. J. B. Goodenough, *J. Solid State Chem.*, **3**, 26 (1974).

33. K. Adachi, K. Sato, and M. Takeda, *J. Phys. Soc. Jpn.*, **26**, 631 (1969).

34. Y. Jeannin and J. Benard, *C. R. Acad. Sci.*, **248**, 2875 (1959).

35. J. Benard and Y. Jeannin, *Adv. Chem. Ser.*, **39**, 191 (1963).

36. L. E. Conroy and K. C. Park, *Inorg. Chem.*, **7**, 459 (1968).

37. A. H. Thompson, K. R. Pisharody, and R. F. Koehler, Jr., *Phys. Rev. Lett.*, **29**, 163 (1972).

38. S. Takeuchi and H. Katsuta, *Nippon Kinzoku Cakkaishi*, **34**, 764 (1970).

39. E. Parthé, D. Hohnke, and F. Hulliger, *Acta Crystallogr.*, **23**, 832 (1967).

40. F. Grønvold, H. Haraldsen, and A. Kjekshus, *Acta Chem. Scand.*, **14**, 1879 (1961).

41. F. Hulliger, *Helv. Phys. Acta*, **33**, 959 (1960).

42. A. Finley, D. Schleich, J. Ackerman, S. Soled, and A. Wold, *Mat. Res. Bull.*, **9**, 1655 (1974).

43. R. G. Dickinson and L. Pauling, J. Am. Chem. Soc., **45**, 1466 (1923).

44. F. Jellinek, G. Brauer, and H. Müller, *Nature (London)*, **185**, 376 (1960).

45. J. A. Wilson and A. D. Yoffe, *Adv. Phys.*, **18**, 193 (1969).

46. R. A. Bromley, R. B. Murray, and A. D. Yoffe, *J. Phys. Chem.*, **5**(7) 759 (1972).

47. P. M. Magie, *Lubric. Eng.*, **22**(7), 262 (1966).

48. H. Shäfer, T. Grofe, and M. Trenksel, *J. Solid State Chem.*, **8**, 14 (1973).

49. A. A. Opalovskii and V. E. Fedorov, *Russian Chem. Rev.*, **35**(3), 186 (1966).

50. R. Huisman, R. DeJonge, C. Haas, and F. Jellinek, *J. Solid State Chem.*, **3**, 56 (1971).

51. J. Rouxel, L. Trichet, P. Chevalier, P. Colombert, and A. Abou Ghaloun, *J. Solid State Chem.*, **29**, 311 (1979).

52. J. C. McMenamin and W. E. Spicer, *Phys. Rev. Lett.*, **29**, 1501 (1972).

53. L. F. Mattheiss, *Phys. Rev. B*, **8**, 3719 (1973).

54. F. M. Williams and F. R. Shepherd, *J. Phys Chem.*, **7**, 4427 (1974).

55. R. S. Title and M. W. Shafer, *Phys. Rev. Lett.*, **28**, 808 (1972).

56. R. S. Title and M. W. Shafer, *Phys. Rev. B*, **8**, 615 (1973).

57. F. Mehran, R. S. Title, and M. W. Shafer, *Solid State Commun.*, **20**, 369 (1976).

58. R. M. M. Fonville, W. Geertsma, and C. Haas, *Phys. Status Solidi B*, **85**, 621 (1978).

59. A. F. Wells, *Structural Inorganic Chemistry*, 5th ed. Clarendon Press, Oxford, 1984, p.33.

60. B. Van Laar, H. M. Rietveld, and D. J. W. Ijdo, *J. Solid State Chem.*, **3**, 154 (1971).

61. R. R. Haering et al., U. S. Patent 4,224,390, September 23, 1980.

62. T. Komatsu and W. K. Hall, *J. Phys. Chem.*, **95**, 9966 (1991).

63. J. M. van den Berg and P. Cossee, *Inorg. Chim. Acta*, **2**, 143 (1968).

64. K. Anzenhofer, J. M. van denBerg, P. Cossee, and J. N. Helle, *J. Phys. Chem. Solids*, **31**, 1057 (1970).

65. R. H. Friend, A. R. Beal, and A. D. Yoffe, *Phil. Mag.*, **35**, 1269 (1977).

66. F. Gamble, F. J. DiSalvo, R. A. Kleem, and T. H. Geballe, *Science*, **168**, 568 (1970).

67. F. R. Gamble, J. H. Osiecki, M. Cais, R. Pisharody, F. J. DiSalvo, and T. H. Geballe, *Science*, **174**, 493 (1971).

68. D.L. Greenway, and R. Nitsche, *J. Phys. Chem. Solids*, **26**, 1445 (1965).

69. G. Said and P. A. Lee, *Phys. Status Solidi A*, **15**, 99 (1973).

70. J. George and C. K. Kumari, *Solid State Commun.*, **49**(1), 103 (1984).

71. Y. Ishizawa and Y. Fugiki, *J. Phys. Soc. Jpn.*, **35**, 1259 (1973).

72. C. R. Whitehouse and A. A. Balchin, *J. Cryst. Growth*, **47**, 203 (1979).

73. B. Palosz, W. Palosz, and S. Gierlotka, *Bull. Mineral.*, **109**, 143 (1986).

74. R. S. Mitchell, Y. Fugiki, and Y. Ishizawa, *J. Cryst. Growth*, **57**, 273 (1982).

75. B. Palosz, *Phys. Status Solidi A*, **80**, 11 (1983).

76. K. Kourtakis, J. DiCarlo, R. Kershaw, K. Dwight, and A. Wold, *J. Solid State Chem.*, **76**, 186 (1988).

77. J. C. Mikkelsen, Jr., *J. Cryst. Growth*, **49**, 253 (1980).

78. A. Weiss and R. Ruthhardt, *Z. Naturforsch. B*, **24**(2), 256 (1969).

79. R. Schöllhorn and A. Weiss, *Z. Naturforsch B*, **27**, 1273 (1972).

80. M. S. Whittingham, *Mat. Res. Bull.*, **9**, 1681 (1974).

81. M. S. Whittingham, *J. Chem. Soc., Chem. Commun.*, **328**(9), (1974).

82. R. Schöllhorn, E. Sick, and A. Lerf, *Mat. Res. Bull.*, **10**, 1005 (1975).

83. M. S. Whittingham, *Materials Science in Energy Technology*, Chap. 9, Academic Press, New York, 1978.

84. M. S. Whittingham, *J. Electrochem. Soc.*, **123**, 315 (1976).

85. R. Chevrel, M. Sergent, and J. Prigent, *J. Solid State Chem.*, **3**, 515 (1971).

86. J. Guillevic, O. Bars, and D. Grandjean, Acta Crystallogr., **32**, 1138 (1976).

87. K. Yvon, A. Paoli, R. Flükiger, and R. Chevrel, *Acta Crystallogr. Sect. B*, **33**, 3066 (1977).

88. O. Fisher, A. Treyvaud, R. Chevrel, and M. Sergent, *Solid State Commun.*, **17**, 721 (1975).

89. R. J. Cava, A. Santoro, and J. M. Tarascon, *J. Solid State Chem.*, **54**, 193 (1984).

90. K. Yvon, *Curr. Top. Mater. Sci.*, **3**, 53 (1979).

91. J. M. Tarascon, F. J. DiSalvo, D. W. Murphy, G. W. Hull, E. A. Rietman, and J. V. Waszczak, *J. Solid State Chem.*, **54**, 204 (1984).

92. R. Schöllhorn, M. Kumpers, A. Lerf, E. Umlauf, and W. S. Schmidt, *Mat. Res. Bull.*, **14**, 1039 (1979).

10

Chalcogenides With the Tetrahedral Structure

The diamond, or tetrahedral structure, is an excellent example of how a simple set of structural relationships can be used to understand the structures of a variety of compounds of increasing complexity. The elements C, Si, Ge, and Sn can form three-dimensional networks in which each atom forms four tetrahedral bonds. The diamond structure of gray tin transforms to white tin at 13.2°C. In white tin, the coordination number is six and the interbond angles are no longer all equal to 109.5°; two are enlarged to 149.5° and four are reduced to 94° (1).

The eight electrons that are required to form four tetrahedral bonds need not be provided solely by these atoms, but can also come from III–V or II–VI pairs of atoms. For example, equal numbers of silicon and carbon (SiC), zinc and sulfur (ZnS), or gallium and phosphorus (GaP) can also provide these electrons for bonding. Furthermore, one-half of the zinc in zincblende (ZnS) can be replaced in an ordered fashion by copper and the remainder by iron. As a result of this ordered arrangement, the repeat unit must now be doubled along one direction. A further replacement of one-half of the iron atoms by tin gives a compound that crystallizes with the stannite structure. Figure 10.1 indicates the structural relationships of a number of phases crystallizing with the tetrahedral structure.

Many semiconductors crystallize with diamond-like structures in which all of the sites are tetrahedrally coordinated. The II–VI semiconductors can accommodate a number of divalent transition metals on these sites. In oxides, several of these species, e.g., Fe(II)($3d^6$) or Ni(II)($3d^8$) prefer octahedral sites. However, in certain chalcogenides, Fe(II) can be stabilized on a tetrahedral site (2–4). It has been shown that Ni(II), which has a much greater octahedral stabilization energy, shows a very limited solubility on tetrahedral sites in II–VI compounds (5). The solubility limits vary among the transition metals in the

Chalcogenides with Tetrahedral Structures

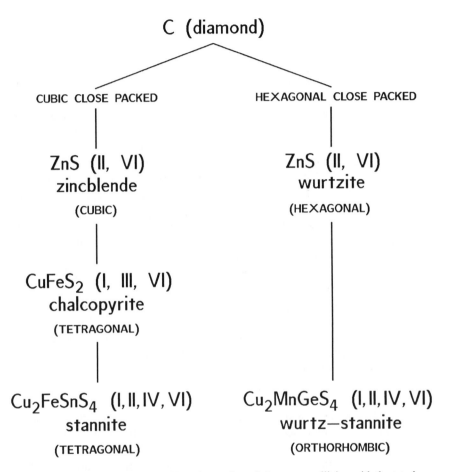

Figure 10.1. Structural relationships of a number of phases crystallizing with the tetrahedral structure.

II–VI semiconductors and is dependent not only on the electron configuration of the transition metal, but also on the structural considerations of the host. For example, in the II-VI compounds, Mn(II) substitutes much more readily than Ni(II) (5).

In this treatment of chalcogenides that crystallize with tetrahedral structures, emphasis is placed on those compounds that illustrate the various structural relationships unique to the chemistry of metal chalcogenides. In addition, some of the unusual magnetic and electronic properties of transition metals in chalco-genides with the tetrahedral structure will be compared to their properties in oxide environments.

A. Zinc Sulfide

Zinc sulfide is probably the most important compound of all the II–VI semicon-ductors. When doped with trace amounts of foreign atoms (Cu, Ag, rare earths, etc.), ZnS can convert different forms of energy into visible light. It has found use in the manufacture of cathode ray tube screens (television screens), photovoltaics, electroluminescence, etc. Many of these applications are dependent on the type and number of defects found in crystals of this material (6–9). However, a principal difficulty in attempts to correlate optical properties to defect concentra-tion has been associated with the problem of obtaining pure single crystals of ZnS. ZnS exists in a number of different crystal modifications. In addition to both the wurtzite and zincblende forms, ZnS shows a large variety of structures derived from these two forms by the insertion of stacking faults. Lendway (10) indicated that crystals with satisfactory structure cannot be obtained by conventional crystal growth methods since both cubic and hexagonal or polytype modifications of ZnS are simultaneously formed in most cases. However, he demonstrated from both experiment and theory that closed transport systems are suitable for growing ZnS crystals. Perfect single ZnS crystals with either hexago-nal or cubic structure with low dislocation density were grown. Ammonium chloride was used as the transport agent and the structure of the crystals was influenced by pulling the sealed silica ampoule through an inhomogeneous tem-perature gradient (9). The growth parameters were chosen so as to maximize the dissolution of faulted nuclei.

Pure zinc sulfide is a diamagnetic semiconductor. The reported magnetic susceptibilities of zincblende and wurtzite are -0.262×10^{-6} and -0.290×10^{-6} emu/g, respectively (11). The band gap for ZnS has been reported to lie between 3.2 and 3.9 eV at 300 K, depending on how the measurement was made (12,13). ZnS was reported to show either n-type or p-type behavior. However, it is generally regarded that p-type ZnS cannot be prepared (14,15). Scott and Barnes (16) suggested that wurtzite is actually sulfur deficient compared to zincblende. Electrical measurements indicated that zinc vacancies were present in zincblende and sulfur vacancies occurred in wurtzite. The entire range in nonstoichiometry was about one atomic percent. Both phases were reported (17) to coexist over a range of temperatures below 1020°C, which was the reported zincblende to wurtzite transition. The difference in free energies of formation for the hexagonal and cubic forms of the II–VI compounds is small, e.g., 229 cal/mol for ZnSe (18), hence it is easy to understand the simultaneous occurrence of the two phases.

The growth of hexagonal ZnS from the melt is consistent with the apparent slow rate of transformation of hexagonal to cubic ZnS (19). There appears, therefore, to be an unresolved controversy regarding the mechanism of the zincblende-wurtzite transformation. A sizable body of data supports the view of Scott and Barnes (16) that nonstoichiometry is the critical factor for the phase transition. Skinner and Barton (20) showed that zincblende can be in equilibrium with either a sulfur-rich or zinc-rich atmosphere. Hence, non-stoichiometry may not be the critical factor for control of wurtzite-zincblende stability.

B. Diamond

Diamond crystallizes with the same structure as the II–VI compounds discussed in this chapter. Despite the fact that it is neither an oxide nor a sulfide, its growing importance in many industrial processes justifies its inclusion in this treatment of solid state chemistry.

The unique properties of diamond, e.g., extreme hardness, high thermal conductivity, high electrical resistivity, and corrosion resistance, have attracted widespread interest for a number of technical applications such as polishing, grinding, cutting, coating on glass lenses, laser windows and lenses, speaker or microphone diaphragms, etc. It may be surprising to discover that diamond is rare in nature since the energy difference between metastable diamond and common graphite is only 453 cal/mol, which is less than RT at room temperature, i.e., 600 cal/mol. This occurrence is primarily because the rate of formation of graphite is much greater than that for diamond. In 1954 (21), scientists at General Electric codeveloped a high-temperature, high-pressure process utilizing a molten nickel catalyst to synthesize diamonds. Diamond is the expected stable phase of carbon at elevated pressure since the density of diamond is higher than that of graphite. It appears that at 1800°C and 70 kbar pressure, nickel dissolves graphite and the less soluble diamond phase crystallizes out. However, the production of bulk diamonds by high-pressure, high-temperature methods is expensive and the diamonds are only of industrial quality. Nevertheless, more than 100,000 lb of synthetic diamonds are produced annually, which find use primarily for cutting, grinding, and polishing operations.

The synthesis of diamond under ambient pressure has been attempted for hundreds of years. In retrospect, since the attempts were carried out under conditions where graphite was the more stable phase, it is not surprising that there was a lack of success (22). Despite considerable doubt by the scientific community concerning the nucleation of diamond under metastable conditions, Percy Bridgman in 1955 (23) made the following observation: "We know from the thermodynamic potential that graphite is ordinarily the preferred form, but this does not enable us to say that the actual precipitate will be graphite and not diamond. As a matter of fact, there are many instances in which an element's unstable form, corresponding to diamond separates from a solidifying liquid or solution in preference to the stable form."

In 1981, Spitsyn et al. (24) reported the following successful synthesis of diamond under ambient pressure:

$$CH_4 + H_2 \xrightarrow[\text{atomic hydrogen}]{\text{heat}} \text{Diamond} + 2H_2$$

The atomic hydrogen could be generated by dissociating molecular hydrogen with hot filaments, flames, or the use of high- and low-pressure RF, microwave, or dc plasmas. The role of atomic hydrogen is to suppress the nucleation and growth of graphitic carbon. In addition, atomic hydrogen saturates and satisfies the carbon bonds at the diamond surface, thereby preventing surface reconstruction. Atomic hydrogen can also remove hydrogen from a hydrogen-covered surface leaving a reactive free radical site that can, in turn, be occupied by a carbon atom in diamond coordination. Hence, diamond growth proceeds in the presence and with the aid of atomic hydrogen.

Diamond films have been prepared at low pressures by microwave plasma chemical vapor deposition (25), induction thermal plasma chemical vapor deposition (26,27), electron-assisted chemical vapor deposition (28), ion beam deposition (29,30), dc arc discharge plasma chemical vapor deposition (31,32), and, finally, hot filament deposition. All the products showed properties similar, to some extent, to those of natural diamond. In all of the techniques, a hydrocarbon, e.g., methane, and hydrogen are dissociated to produce atomic hydrogen and hydrocarbon radicals.

In the preparation of diamond films, the density of nucleation can be increased by prescratching the substrate surface with diamond particles (33). This scratching process has been used to achieve complete film coverage of the substrate when the rate of deposition is low. High rate processes result in films that are rough and are composed of large grains. The quality of all films produced at low pressure is poor because, during the initial stages of diamond growth, numerous nuclei are produced that are randomly oriented. In addition, if a hot filament reactor is used, graphite, hydrogen, and metal atoms are incorporated into the diamond film. The impurities are found at the grain boundaries of the growing nuclei.

Badzian and Badzian (33) indicated that a buffer layer of βSiC (or other carbides) improves diamond films on various substrates. However, Niu et al. (34) indicated that the formation of continuous films on diamond-scratched substrates of Cu and Au indicates that carbide formation is not necessary to initiate nucleation sites for diamond. Neither copper nor gold is reported to form stable carbides. However, the adhesion of diamond films to silicon substrates is undoubtedly due to strong bonding via a carbide intermediate. Moreover, the adhesion of films to

Cu and Au substrates is poor, and this is consistent with their inability to form carbide phases.

C. Chalcopyrite and Related Structures

a. Chalcopyrite

Burdick and Ellis (35,36) were the first to investigate the structure of $CuFeS_2$. They showed that the atoms were positioned so as to form a face-centered tetragonal structure. Alternate rows of copper and iron atoms were found to order in planes perpendicular to the tetragonal axis.

Pauling and Brockway (37) reexamined the structure of chalcopyrite using a natural single crystal and determined that this material crystallizes in the $I\bar{4}2d$ space group. The structure of chalcopyrite was found to be similar to that of cubic ZnS (Fig. 10.2). In the ZnS structure, each zinc atom is tetrahedrally coordinated by four sulfur atoms, and each sulfur atom is tetrahedrally coordinated by four zinc atoms. The sulfur atoms are positioned so that they form a face-centered array. In the chalcopyrite structure (Fig. 10.3), the zinc atoms in zincblende are replaced by an equal number of copper and iron atoms. The bonding in $CuFeS_2$ is such that each metal atom is approximately tetrahedrally coordinated to four sulfur atoms and each sulfur atom is similarly coordinated to a pair of copper atoms and a pair of iron atoms, respectively. The copper and iron atoms alternate along one direction, that is assigned the c axis. This results in a tetragonal cell with a c/a value close to 2.

In 1958, Donnay et al. (38) confirmed the results of Pauling and Brockway by both X-ray diffraction and neutron diffraction studies carried out on natural crystals of $CuFeS_2$. The magnetic unit cell was found to be identical to the chemical unit cell. The material was found to be ordered antiferromagnetically at room temperature. The antiferromagnetic structure consists of an arrangement in which two iron atoms connected to a common sulfur atom have oppositely directed moments (Fig. 10.4). The iron atoms are ordered ferromagnetically in (001) layers that are coupled antiferromagnetcally along the c crystallographic direction. This magnetic structure is equivalent to face-centered ordering of the third kind but with one-half of the metal atoms (Cu) having zero moment. Ordering of the third kind is indicative of more than one type of nearest-neighbor or next nearest-neighbor interaction (39). For $CuFeS_2$ these interactions can result in a partial participation of the $3d$ electrons of iron in covalent bonding with the sulfur atoms. This covalent bonding has been used to explain the observed moment, obtained from neutron diffraction data, of 3.85 μ_B that is less than the expected value of 5.0 μ_B for Fe(III).

In 1961, Teranishi (40) confirmed the findings of Donnay et al. (38), indicating chalcopyrite to be an antiferromagnetic substance with a Néel point of 823 K.

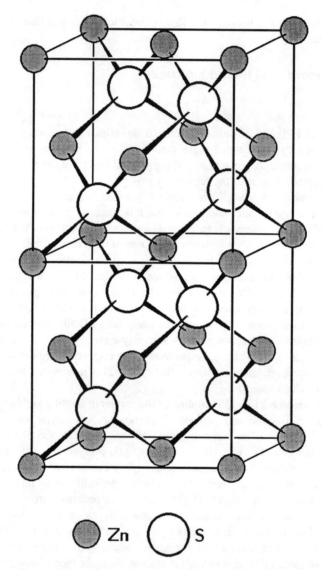

Figure 10.2. Two unit cells of the cubic zincblende form of ZnS.

Mössbauer studies have been reported on natural samples of chalcopyrite (41,42). Spectra obtained at 298 K showed a six line magnetic hyperfine splitting, indicating the presence of magnetic order. An internal field of 354 kOe and an isomer shift of 0.49 mm/sec (with respect to the standard sodium iron(III) nitropentacyanide dihydrate at 298 K) have been reported. These values of internal field

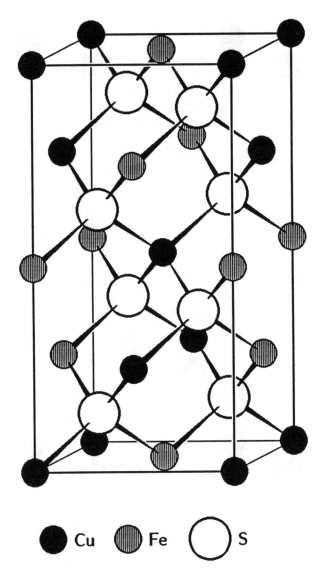

Figure 10.3. The chalcopyrite structure.

strength and isomer shift are less than those that are observed for iron (d^5) oxides, but are consistent with data obtained for other iron (d^5) sulfide systems (43).

Adams et al. (44) have prepared and characterized samples of $CuFeS_2$. A polycrystalline sample of chalcopyrite was heated in a dynamic vacuum of 10^{-5} torr and the weight of the sample was monitored as a function of temperature. $CuFeS_2$ was shown to decompose at 663 K. This temperature is substantially

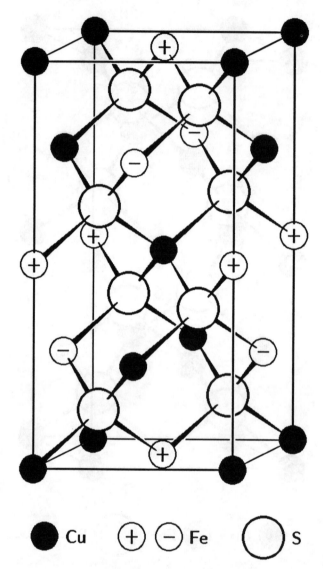

Cu ⊕ ⊖ **Fe** ◯ **S**

Figure 10.4. Magnetic ordering in chalcopyrite.

below the Néel point of 823 K. The Néel point was confirmed by DiGuiseppe et al. (45) by extrapolating Mössbauer data from samples of the system $CuGa_{1-x}Fe_xS_2$, which did not decompose below the antiferromagnetic transition. In the system $CuGa_{1-x}Fe_xS_2$, several interesting observations were made (45) as to the magnetic behavior of these compositions with increasing iron content. Compositions with $x < 0.3$ behave as a dilute magnetic system with the magnetic properties determined by the number of nearest-neighbor iron ions. Increased iron concentration results in an increased number of nearest-neighbor iron ions, thus increasing the antiferromagnetic interactions through the formation of Fe–Fe pairs. As the number of these pairs increases, iron ion clusters occur resulting in deviations from Curie–Weiss behavior for compositions with $x \geqq 0.3$.

Single crystals of $CuFeS_2$ were grown from the melt by means of a modified Bridgman technique (44). The principal difficulty in growing the crystals is the loss of sulfur above 390°C. Hence, the composition of the final product tends to be $CuFeS_{2-x}$. This problem was overcome by annealing the boules with excess sulfur in sealed evacuated silica tubes. The amount of sulfur to be added was determined from careful density and X-ray measurements. Single crystals were grown that were of superior quality to those found in nature.

b. Solid Solutions of $(ZnS)_{1-x}(CuMS_2)_x$ $(M = Al, Ga, In, or Fe)$

There have been relatively few studies (45–49) carried out on the preparation and characterization of solid solutions formed between ZnS and $CuMS_2$ (M = Al, Ga, In, or Fe). Apple (46) and Robbins and Miksovsky (47) have investigated the extent of the solid solutions $(ZnS)_{1-x}(CuMS_2)_x$ (M = Al, Ga, or In) and have determined their optical properties. It was noted that the ternary chalcopyrites, $CuMS_2$ (M = Al, Ga, or In), were totally miscible with ZnS. In the $(ZnS)_{1-x}(CuMS_2)_x$ (M = Al, Ga, or In) systems, the cubic zincblende structure was the stable structure for substitution of up to 30 mol% $CuAlS_2$, 40 mol% $CuGaS_2$ or 50 mol% $CuInS_2$, respectively. Moh (48) and Kajima and Sugaki (49) reported the phase relations between ZnS and $CuFeS_2$ above 300°C. In the pseudobinary $ZnS - CuFeS_2$ system, the maximum $CuFeS_2$ in ZnS is approximately 1.6 mol% at 800°C.

ZnS is used as an IR window material because of its wide transmission range in the far infrared. However, ZnS is soft, which limits its suitability for some applications. Single crystals of members of the system $(ZnS)_{1-x}(CuGaS_2)_x$ (x=0.053, 0.103) showed good transmission in the far-infrared range and were much harder than ZnS (50).

$CuAlS_2$ and $CuInS_2$ crystallize with the same structure as $CuGaS_2$ and they show similar chemical and physical properties. Therefore, the systems $(ZnS)_{1-x}(CuMS_2)_x$ (M = Al or In) have been analyzed similarly to the $(ZnS)_{1-x}(CuGaS_2)_x$ system (50). Polycrystalline samples of $(ZnS)_{1-x}(CuMS_2)_x$ (M = Al or In) where $x \leqq 0.3$ were prepared directly from the elements (51). X-Ray diffraction analyses indicated that $(ZnS)_{1-x}(CuMS_2)_x$ (M = Al or In) polycrystalline samples were all single phase with the cubic zincblende structure. These results are in good agreement with previous studies (46,47), that have reported the extent of the solid solutions $(ZnS)_{1-x}(CuMS_2)_x$ (M = Al, Ga, or In). Because of similarities in structure, unit cell dimensions, and bond type, the ternary sulfides $CuMS_2$ (M = Al, Ga, or In) were found to be totally miscible with ZnS.

The cell parameters of the polycrystalline samples are plotted as a function of chalcopyrite concentration for $(ZnS)_{1-x}(CuMS_2)_x$ (M = Al or In) in Fig. 10.5. At chalcopyrite concentrations less than 30.0 mol%, the cell parameters decrease linearly with increasing amounts of substituted chalcopyrite in the aluminum system, and increase linearly in the indium system in accordance with Vegard's law. The cell parameter data are in good agreement with those reported by previous papers (47,52). The composition of unanalyzed single crystals, which are grown by chemical vapor transport, can be obtained from the linear relationship between the cell parameter and the concentration of substituted chalcopyrite.

Single crystals of $(ZnS)_{1-x}(CuMS_2)_x$ (M = Al or In)(x = 0.05 or 0.10) were grown by chemical vapor transport using iodine as the transport agent. All single crystals grown crystallized with the cubic zincblende structure. The properties of these compounds are summarized in Table 10.1 (51). The compositions of $(ZnS)_{1-x}(CuMS_2)_x$ (M = Al or In) single crystals were determined by comparing their cell parameters with those obtained from a plot of cell parameter vs composition for the standard polycrystalline materials (Fig. 10.5). The results of these determinations for $(ZnS)_{1-x}(CuMS_2)_x$ (M = Al or In) are given in Table 10.1. It can be seen that the actual composition of some of the transported crystals deviated from their nominal composition. Hence, the determination of composition by comparison of single crystal cell parameters with those of known standards is essential.

The hardness values, as determined by the Knoop method, are also given in Table 10.1. The measured hardness of pure ZnS is 153 kg/mm^2, which is in good agreement with previous investigations (54-56). It is noted that the hardness values of the crystals containing chalcopyrite substitution show a significant increase compared to the value of the pure end member. In previous studies (57,58), it has been reported that chalcopyrites (I–III–VI$_2$) are much harder than II–VI compounds and that the hardness of chalcopyrites decreases from $CuAlS_2$ to $CuInS_2$. The substitution of $CuAlS_2$ results in a relatively larger increase in the hardness of ZnS than an equivalent substitution of $CuGaS_2$ or $CuInS_2$. These results are in agreement with those reported by He et al. (57) in which the measured value of the hardness of $CuAlS_2$ is greater than that of $CuGaS_2$. Shay and Wernick (56) speculated that as the atomic number increases in the same family, atoms are more polarizable and hence a decrease in the measured hardness would be anticipated. Even at the low concentrations of chalcopyrite substituted for ZnS in this study, the effective increase in the hardening by $CuAlS_2$ can be observed (Table 10.1).

The IR transmission data are also summarized in Table 10.1. The results indicate that pure ZnS transmits in the range of 1.5–14 μm, which is in good agreement with previous reports (53–55). $CuAlS_2$ causes a marked decrease in the upper end of the transmission of ZnS and $CuGaS_2$ does narrow the transmission

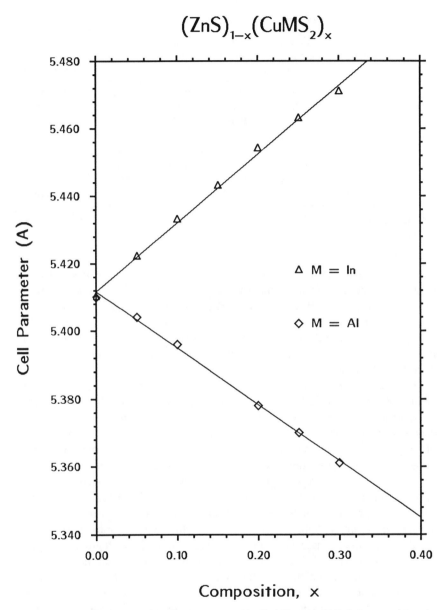

Figure 10.5. Variation of cell parameter with CuAlS$_2$ and CuInS$_2$ composition in (ZnS)$_{1-x}$(CuMS$_2$)$_x$ (M = Al or In)

Table 10.1. Properties of $(ZnS)_{1-x}(CuMS_2)_x$ (M = Al, Ga, In, or Fe) Single Crystals

	Nominal composition (x)	Crystal composition (x)	Cell parameter (Å)	Knoop hardness (Kg/mm²)	Decomposition temperature (°C)	IR Transmission range (μm)
	0.0	0.0	5.410(2)	153(10)	570	2.5–14
M = Al	0.05	0.027	5.407(2)	206(20)	580	4.0–10
	0.10	0.069	5.400(2)	275(35)	595	4.0–10
M = Ga	0.05	0.053	5.403(2)	250(10)	670	4.5–13
	0.10	0.103	5.397(2)	298(20)	680	4.5–13
M = In	0.05	0.051	5.422(2)	211(20)	480	—[a]
	0.10	0.105	5.433(2)	254(20)	450	—[a]
M = Fe	0.10	0.02	5.412(2)	226(15)	530	5.0–14

[a] The transmission of single crystals with 0.01 cm thickness is less than 30% in the range of 2.5–50 μm.

range, particularly at the low end. However, these materials still show good transmission in the long wavelength IR range. Between 2.5 and 50 μm, $CuInS_2$ reduces the magnitude of IR transmission by less than 30%. The thermal stability data, taken in a flowing oxygen atmosphere, show that the chalcopyrites $CuAlS_2$ and $CuGaS_2$ increase the decomposition temperature of pure ZnS (Table 10.1). For the ZnS–$CuInS_2$ systems, the onset temperature of decomposition is lower than that of pure ZnS.

The electrical and optical properties of the $CuMS_2$ compounds (M = Al, Ga, or In) have been reported by Tell et al. (58). In early studies, Shay et al. (59–61) indicated that many of the electronic properties can be explained if the valence band is assigned considerable d-character. This can result from the hybridization of copper 3d orbitals and anion sp states. In a more recent paper, Jaffe and Zunger (62) calculated the electronic structure of these ternary chalcopyrite semiconductors. They reported that almost all of the copper d-electrons occupy the upper valence band and there are empty group III metal states in the conduction band. The electronic structure of the upper valence band consists primarily of copper 3d–sulfur 3p hybrid wave functions which interact most strongly for $CuAlS_2$. The empty conduction band is composed of unoccupied group III–group VI anion states. The filled group III–group VI anion valence band lies below the copper 3d–sulfur 3p band. It is, therefore, not surprising that $CuFeS_2$ shows different physical and chemical properties from the other $CuMS_2$ (M = Al, Ga, or In) I–III–VI_2 compounds. $CuFeS_2$, chalcopyrite, is the only compound that contains a magnetic ion among these tetrahedrally coordinated semiconductors. With respect to the measured electrical properties, there is little difference between chalcopyrite and the nonmagnetic analogs. However, the observed Hall mobility for chalcopyrite of 35 cm²/V-sec at 80 K (61) is small, which suggests some effect of the unpaired iron 3d electrons on the electrical properties. Furthermore, from optical measurements, Goodman and Douglas (63) and Austin et al. (64) reported that the absorption edge of 0.5 eV for $CuFeS_2$ is much smaller than the value of 2.5 eV measured for $CuGaS_2$. The difference in the absorption edge of $CuFeS_2$ may also be related to the delocalization of Fe(III) $3d$ electrons.

The magnetic properties of $CuFeS_2$ show the effect much more clearly. Neutron diffraction (38) and static magnetic measurements (61) have shown that $CuFeS_2$ is antiferromagnetic with $T_N = 550°C$. The effective magnetic moment associated with $Fe(3d^5)$ was only 3.85 μ_B. This moment cannot be interpreted with an assignment of Fe(III) $3d^5$ that would be consistent with Mössbauer studies, which assign iron as a trivalent species (41,65). Finally, it was shown by Sato and Teranishi (66) that for the systems $CuFe_xAl_{1-x}S_2$ and $CuFe_xGa_{1-x}S_2$ the iron $3d$ electrons are localized when the iron concentration is small, but undergo a transition to the delocalized state at a critical value of x. These results are consistent with those obtained by Sainctavit et al. (67) from XANES spectra. They reported that for $CuFeS_2$ there is additional strong hybridization of anion $3p$ and delocalized iron $3d$ states.

For the system $(ZnS)_{1-x}(CuFeS_2)_x$ where $0.025 \leq x \leq 0.3$, X-ray diffraction patterns of polycrystalline samples indicated that there were two phases present, namely, the cubic zincblende structure and the tetragonal chalcopyrite structure. This is consistent with the report by Moh (48) that the maximum solubility of $CuFeS_2$ in ZnS was approximately 1.6 mol% at 800°C. Undoubtedly, the narrow solubility limit of $CuFeS_2$ in ZnS is related to their differences in bonding. For $CuFeS_2$, there is at least a partial participation of the iron unpaired $3d$ electrons with the uppermost sulfur valence bands. This is also consistent with the evidence previously discussed that supports the delocalization of $3d$ electrons over the bonding bands. Furthermore, the cell dimensions of $CuFeS_2$ (a = 5.29 Å and c = 10.43 Å) are smaller than those of $CuGaS_2$ (a = 5.36 Å and c = 10.49 Å) even though the radius of Fe(III) (r = 0.49 Å) is larger than that of Ga(III) (r = 0.47 Å). This is consistent with the concept of increased metal-sulfur bonding in $CuFeS_2$ resulting from partial delocalization of Fe(III) $3d$ electrons and admixture with anion p-states.

The composition of $(ZnS)_{0.98}(CuFeS_2)_{0.02}$ single crystals grown by chemical vapor transport was determined by magnetic measurements (51). The phase crystallized with the zincblende structure. Magnetic susceptibility measurements were made as functions of both field and temperature. Two separate crystals were measured and showed paramagnetic behavior without any field dependency at either room temperature or at liquid nitrogen temperature. The reciprocal magnetic susceptibility of $(ZnS)_{0.98}(CuFeS_2)_{0.02}$ is plotted versus temperature in Fig. 10.6 and shows Curie–Weiss behavior with a Weiss constant of -110 K. The composition was calculated by comparing the observed Curie constant with that calculated for a spin-only moment (5.9 μ_B) of the Fe(III) ions (Table 10.1). It can be seen that crystals grown from the charge containing $CuFeS_2$ to ZnS ratio of 10/90 gave a $CuFeS_2$ content of 2 mol%. The apparent difference between this

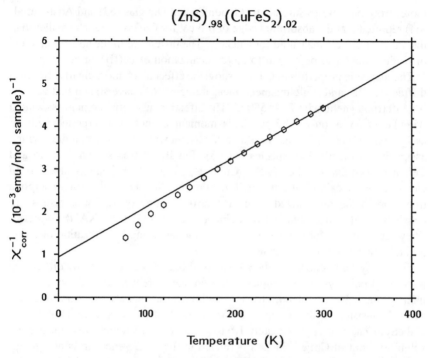

Figure 10.6. Variation of magnetic susceptibility with temperature for a single crystal of $(ZnS)_{0.98}(CuFeS_2)_{0.02}$.

solubility of $CuFeS_2$ in ZnS (51) and that reported by Moh (48) is undoubtedly due to the formation of iron clusters. This is consistent with the observation of an appreciable Weiss constant (51).

The properties of $(ZnS)_{0.98}(CuFeS_2)_{0.02}$ single crystals are summarized in Table 10.1. These crystals give the same IR transmission as ZnS at the long wavelength end, but there appears to be a cut off below 5.0 μm. The microhardness and thermal stability data show that $CuFeS_2$ increases the hardness of pure ZnS and decreases the decomposition temperature.

c. Stannite and Wurtz-Stannite

The structure of the mineral stannite Cu_2FeSnS_4 was first determined by Brockway (68) who showed that it was closely related to that of chalcopyrite. The structure is shown in Fig. 10.7. It can be seen that there are alternate layers perpendicular to the 4-fold axis, which contain either copper atoms or an equal number of iron and tin atoms. The space group and all cell dimensions for this compound as well as that for a number of related materials are given in Table 10.2.

Nitsche et al.(69) and Schäfer and Nitsche (70) prepared single crystals of the

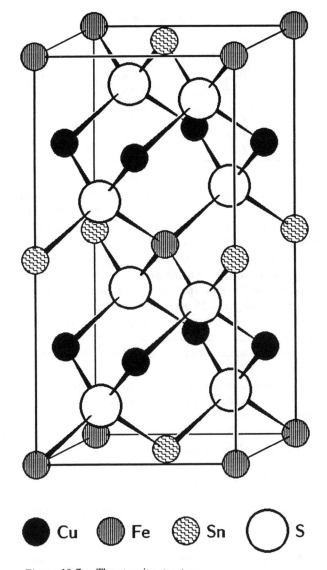

Figure 10.7. The stannite structure.

compounds listed in Table 10.2 and the *I4̄2m* and *Pmn2₁* structures are shown in Figs. 10.8 and 10.9. Both these structures are characterized by nearly tetrahedral coordination about both the metal and chalcogen atoms. The *I4̄2m* structure is based on a double zincblende cell similar to the chalcopyrite structure, but with an additional ordering of the group IIB and group IVB atoms. The *Pmn2₁* structure is based on a doubled wurtzite cell (Fig. 10.10). The compound briartite,

Table 10.2. Iron $3d^6$ Compounds of the Type $CuFe_{0.5}A_{0.5}X_2$ Where A = Si, Ge, Sn; X = S, Se

Compound	Space group	a_0 (Å)	b_0 (Å)	c_0 (Å)
$CuFe_{0.5}Si_{0.5}S_2$	$Pmn2_1$	7.404	6.411	6.140
$CuFe_{0.5}Ge_{0.5}S_2$	$I\bar{4}2m$	5.330		10.528
$CuFe_{0.5}Sn_{0.5}S_2$	$I\bar{4}2m$	5.46		10.725
$CuFe_{0.5}Si_{0.5}Se_2$	$I\bar{4}2m$	5.549		10.951
$CuFe_{0.5}Ge_{0.5}Se_2$	$I\bar{4}2m$	5.590		11.072
$CuFe_{0.5}Sn_{0.5}Se_2$	$I\bar{4}2m$	5.664		11.33

$CuFe_{0.5}Ge_{0.5}S_2$, which crystallizes with the zincblende-related structure, has been characterized magnetically by Allemand and Wintenberger (71). From room temperature to 12 K, the compound showed paramagnetic behavior. The observed magnetic moment was 4.92 μ_B that agrees with the theoretical value of 4.91 μ_B calculated for an iron $3d^6$ (e^3, t_2^3; S=2) atom. The Mössbauer spectra of $CuFe_{0.5}Ge_{0.5}S_2$ were obtained from room temperature to 4 K by Imbert et al. (72). The reported isomer shift with respect to sodium nitroprusside is 0.85 mm/sec with a quadrupole splitting of 2.56 mm/sec. A six line hyperfine splitting occurs at 12.3 K and below, where briartite undergoes antiferromagnetic ordering. The low Néel temperature, 0.85 mm/sec isomer shift, 2.56 mm/sec quadrupole splitting, and spin-only moment are features associated with iron $3d^6$ in a tetrahedral environment.

The similarity between the briartite and chalcopyrite structures suggested that a solid solution series could be formed between these two materials. Since briartite contains only iron $3d^6$ and chalcopyrite $3d^5$, the intermediate members of this series contain both $3d^6$ and $3d^5$ iron. The amount of iron $3d^6$ and iron $3d^5$ present in any member of this series is given by

$$CuFe_xGe_{1-x}S_2 = CuFe(3d^5)_{2x-1}Fe(3d^6)_{1-x}Ge_{1-x}S_2 \ (0.5 \leqq x \leqq 1)$$

Members of the solid solution series between chalcopyrite and briartite were grown by chemical vapor transport (73). X-Ray studies of these crystals indicated that they adopt the space group $I\bar{4}2m$. The presence of iron $3d^5$ and $3d^6$ was confirmed by Mössbauer spectra when $x \leqq 0.75$. For samples where $x \geqq 0.85$, there was evidence of magnetic order. Magnetic susceptibility measurements of the sample where $x = 0.53$ indicated that Curie–Weiss behavior is obeyed for this composition. When x increases, deviations from the Curie–Weiss law occur and complex magnetic ordering results.

The quaternary chalcogenide Cu_2ZnGeS_4 also crystallizes with the stannite structure but the compounds $Cu_2ZnGeSe_4$, Cu_2ZnSiS_4, and $Cu_2ZnSiSe_4$ belong to the orthorhombic superstructure of wurtzite called wurtz-stannite, space group $Pmn2_1$ (Fig. 10.9). Single crystals of these quaternary chalcogenides were prepared by vapor transport of stoichiometric amounts of the elements with iodine

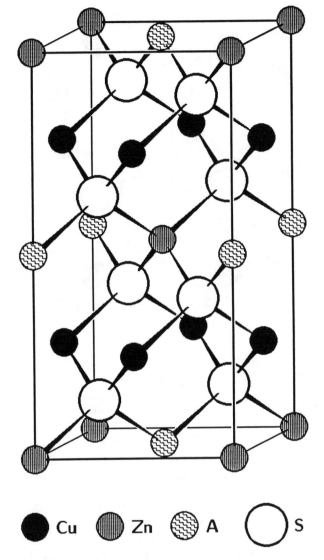

Figure 10.8. Tetragonal *I̅42m* structure of stannite.

as the transport agent (74). Table 10.3 summarizes the results of the X-ray, density, electrical, and optical measurements. The observed changes in the optical bandgap are in agreement with what would be expected in substituting Ge for Si or Se for S. The discrepancy between the optical band gap and the electrical band gap obtained from $\sigma = \sigma_0 \exp(-E_g/2kT)$ is undoubtedly due to the extrinsic nature of these compounds.

In this class of structurally related compounds represented by $Cu(I)_2 B(II)$

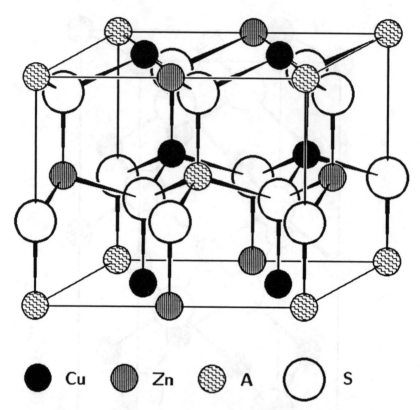

● Cu   ◍ Zn   ⊕ A   ◯ S

Figure 10.9. The orthorhombic $Pmn2_1$ structure of wurtz-stannite.

C(IV)X$_4$, where B = Mn, Fe, Co, Ni, Zn, Cd, Hg, C = Si, Ge, Sn, and X = S, Se (54), most of the compounds crystallize with superstructures of the zinc-blende or wurtzite type (Figs. 10.8 and 10.9). This gives the orthorhombic wurtz-stannite or the tetragonal stannite structure. The stannite structure is obtained by a doubling of the II–VI zincblende cell and replacing the divalent cation with equal amounts of a monovalent and trivalent species as in the mineral chalcopyrite Cu(I)Fe(III)S$_2$ (Fig. 10.3). If in the chalcopyrite cell the trivalent species are replaced with equal amounts of divalent and tetravalent cations, then the quaternary stannite Cu$_2$B(II)C(IV)X$_4$ with a tetragonal cell is obtained. The same type of structural relationship exists between the II-VI wurtzite unit cell and the wurtz-stannite cell. The wurtz-stannite structure is based on a hexagonal close-packed arrangement of the anions as exists in the wurtzite structure (Fig. 10.10). The wurtz-stannite structure crystallizes with an hexagonal cell but can be described as an orthorhombic cell (70). This class of compounds is unique among tetrahedrally coordinated semiconductors in their ability to incorporate a sizable mole fraction of several transition metals.

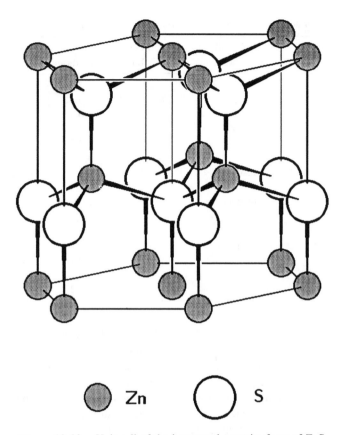

Zn ● ○ S

Figure 10.10. Unit cell of the hexagonal wurtzite form of ZnS.

In these quaternaries, the cations in both the stannite and the wurtz-stannite structures are ordered. Planes of divalent and tetravalent cations are separated by planes containing only copper(I). In these ordered structures, two divalent cations are never bonded to the same anion as they are in the II–VI compounds. Therefore, the antiferromagnetic superexchange interactions are considerably weaker as the transition metal ions are further apart. This allows for a transition to ferromagnetic alignment. Consequently, saturation magnetization can be achieved at relatively low fields (70). This class of compounds is interesting for a number of reasons. They are unique among the diamond-like structures in their ability to incorporate a large mole fraction of several transition metals on tetrahedral sites. Even ions that are not readily stabilized on tetrahedral sites, e.g., Ni(II), can occupy the B-site in these quaternaries.

The compound Cu_2ZnGeS_4 is unique in that it crystallizes in the tetragonal structure at temperatures below 810°C and the wurtz-stannite structure above 810°C (76). The compound Cu_2FeGeS_4 crystallizes with the tetragonal structure

Table 10.3. Physical Measurements of Quaternary Chalcogenides

Lattice	Cu_2ZnSiS_4 Orthorhombic	$Cu_2ZnSiSe_4$ Orthorhombic	Cu_2ZnGeS_4 Orthorhombic	$Cu_2ZnGeSe_4$ Tetragonal
a_0 (Å)	7.44	7.83	7.50	5.61
b_0 (Å)	6.39	6.73	6.48	5.61
c_0 (Å)	6.13	6.44	6.18	11.05
$\rho_{calc.}$ (g cm^{-3})	3.97	5.25	4.35	5.54
$\rho_{obs.}$ (g cm^{-3})	3.94	5.22	4.37	5.50
RT resistivity(Ω cm)	Insulator	2×10^3	1	2×10^{-3}
Liquid N_2 resistivity (Ω cm)	Insulator	5×10^6	2×10^1	4×10^{-3}
Electrical activation energy (eV)	Insulator	0.3	0.03	0.01
Seebeck Sign	Immeasurable	plus	plus	plus
Optical band edge (eV)	3.25	2.33	2.1	1.29

for all temperatures. Doverspike et al. (77) prepared orthorhombic samples of $Cu_2Zn_{1-x}Fe_xGeS_4$ where $0 \leqq x \leqq 0.15$ by rapid quenching from 900°C. At $x = 0.20$, a small amount of the tetragonal phase is always present in the quenched sample. The phase transformation from tetragonal to orthorhombic could be reversed by either annealing the orthorhombic phase below the transition point or by applying elevated pressure. Magnetic susceptibility measurements indicate that there is no significant variation in the Curie or Weiss constants obtained from the tetragonal or orthorhombic phases. Hence, the environment of the $3d^6$ iron is essentially equivalent in the two structures.

References

1. A. F. Wells, *Structural Inorganic Chemistry*, Oxford University Press, Oxford, 5th ed. 1984, p.122.

2. R. Pappalardo and R. E. Dietz, *Phys. Rev.*, **123**, 1188 (1961).

3. K. Smith, J. Marsello, R. Kershaw, K. Dwight, and A. Wold, *Mat. Res. Bull.*, **23**, 1423 (1988).

4. H. Swagten, A. Twardowski, and W. de Jonge, *Phys. Rev. B*, **39**(4) 2568 (1989).

5. G. K. Czamanski and F. E. Goff, *Econ. Geol.*, **68**, 258 (1973).

6. J. Tauc, *J. Phys. Chem. Solids*, **11**, 345 (1959).

7. G. F. Neumark, *Phys. Rev.*, **125**, 838 (1962).

8. H. Haupt and H. Nelkowski, *Z. Naturfosch. A*, **24**, 1904 (1969).

9. E. Lendway, *J. Cryst. Growth*, **10**(1), 77 (1971).

10. E. Lendway, *Acta Techniea. Academiae, Scientiarum Inorganeae*, **80**, 451 (1975).

11. S. Larach and J. Turkevich, *Phys. Rev.*, **98**, 1015 (1955).

12. G. Cheroff and S. P. Keller, *Phys. Rev.*, **111**, 98 (1958).

13. M. V. Fok, *Sov. Phys. Solid State*, **5**, 1085 (1963).

14. F. F. Morehead, *J. Electrochem. Soc.*, **110**, 285 (1963).

15. F. F. Morehead and A.B. Fowler, *J. Electrochem. Soc.*, **109**, 688 (1962).

16. S. D. Scott and H. L. Barnes, *Can. Mineral.*, **9**, 306 (1967) Abstr.

17. S. D. Scott and H. L. Barnes, *Geochim. Cosmochim. Acta*, **36**, 1275 (1972).

18. L. Cemic and A. Neuhaus, *High Temp. High Pressure*, **6**, 203 (1974).

19. B. J. Fitzpatrick, *J. Cryst. Growth*, **86**, 106 (1988).

20. B. J. Skinner and P. B. Barton, *J. Am. Mineral.*, **45**, 612 (1960).

21. F. P. Bundy, N. T. Hall, H. M. Strong and R. J. Wentorf, Jr., *Nature (London)*, **176**. 51 (1955).

22. G. Davies, *Diamond*, Adam Hilger, Bristol, 1984.

23. P. W. Bridgman, *Sci. Am.*, **193**, 42 (1955).

24. B. V. Spitsyn, L. L. Bouilov, and B. V. Deryaguin, *J. Cryst. Growth*, **52**, 219 (1981).

25. M. Kamo, Y. Sato, S. Matsumot, and N. Setaka, *J. Cryst. Growth*, **62**, 642 (1983).

26. S. Matsumoto, *J. Mater. Sci. Lett.*, **4**, 600 (1985).

27. S. Matsumoto, M. Hino, and T. Kobayashi, *Appl. Phys. Lett.*, **51**, 737 (1987).

28. A. Sawabe and T. Inuzuka, *Appl. Phys. Lett.*, **46**, 146 (1985).

29. E. G. Spencer, P. H. Schmidt, D. C. Joy, and F. J. Sansalone, *Appl. Phys. Lett.*, **29**, 118 (1976).

30. K. Suzuki, A. Swabe, H. Yasuda, and T. Inuzuka, *Appl. Phys. Lett.*, **50**, 728 (1987).

31. K. Kurihara, K. Sasaki, M. Kawarada, and N. Koshino, *Appl. Phys. Lett.*, **52**, 437 (1988).

32. F. Akatsuka, Y. Hirose, and K. Komaki, *Jap. J. Appl. Phys. Lett.*, **27**(9), L1600 (1988).

33. A. R. Badzian and T. Badzian, *Surf. Coat. Technol.* **36**, 283 (1988).

34. C-M. Niu, G. Tsagaropoulos, J. Baglio, K. Dwight, and A. Wold, *J. Solid State Chem.*, **91**, 47 (1991).

35. C. L. Burdick and J. H. Ellis, *Proc. Natl. Acad. Sci. U.S.A.*, **3**, 644 (1917).

36. C. L. Burdick and J. H. Ellis, *J.A.C.S.*, **39**, 2518 (1917).

37. L. Pauling and L. O. Brockway, *Z. Krist.*, **82**, 188 (1932).

38. G. Donnay, L. M. Corliss, J. D. H. Donnay, N. Elliot, and J. Hastings, *Phys. Rev.*, **112**, 1917 (1958).

39. J. B. Goodenough, *Magnetism and the Chemical Bond*, Interscience Monographs on Chemistry, Inorganic Chemistry Section, Vol. 1, F. A. Cotton (ed.). Interscience, John Wiley, New York, 1963.

40. T. Teranishi, *J. Phys. Soc. Jpn.*, **16**, 1881 (1961).

41. E. Frank, *Il. Nuovo Cimento*, **58B**, 407 (1968).

42. N. N. Greenwood and H. J. Whitfield, *J. Chem. Soc. (A)*, 1697 (1968).

43. N. N. Greenwood and T. C. Gibb, *Mössbauer Spectroscopy*, Chapman and Hall, London, 1971, pp. 238-286.

44. R. Adams, P. Russo, R. Arnott, and A. Wold, *Mat. Res. Bull.*, 7, 93 (1972).

45. M. DiGiuseppe, J. Steger, A. Wold, and E. Kostiner, *Inorg. Chem.*, **13**, 1828 (1974).

46. E. F. Apple, *J. Electrochem. Soc.*, **105**, 251 (1958).

47. M. Robbins and M. A. Miksovsky, *J. Solid State Chem.*, 5, 462 (1972).

48. G. H. Moh, *Chem. Erde.*, **33**(3), (1974).

49. S. Kojima and A. Sugaki, *Mineral. J.*, **12**, 15 (1984).

50. Y. R. Do, R. Kershaw, K. Dwight, and A. Wold, *J. Solid State Chem.*, **96**, 360 (1992).

51. Y. R. Do, K. Dwight, and A. Wold, *J. Mat. Chem.*, **4**, 1014 (1992).

52. P. C. Donohue and P. E. Bierstedt, *J. Electrochem. Soc.*, **121**, 327 (1974).

53. P. Wu, R. Kershaw, K. Dwight, and A. Wold, *Mat. Res. Bull.*, **24**, 49 (1989).

54. J. DiCarlo, M. Albert, K. Dwight, and A. Wold, *J. Solid State Chem.*, **87**, 443 (1990).

55. C-M. Niu, R. Kershaw, K. Dwight, and A. Wold, *J. Solid State Chem.*, **85**, 262 (1990).

56. J. L. Shay and J. H. Wernick, *Ternary Chalcopyrite Semiconductors: Growth, Electronic Properties and Applications*. Pergamon Press, New York 1975.

57. X-C. He, H-S. Shen, P. Wu, K. Dwight, and A. Wold, *Mat. Res. Bull.*, **23**, 799 (1988).

58. B. Tell, J. L. Shay, and H. M. Kasper, *J. Appl. Phys.*, **43**, 2469 (1972).

59. J. L. Shay and H. M. Kasper, *Phys. Rev. Lett.*, **29**, 1162 (1972).

60. J. L. Shay, B. Tell, H. M. Kasper, and L. M. Schiavone, *Phys. Rev. B*, **5**, 5003 (1972).

61. J. L. Shay, B. Tell, H. M. Kasper, and L. M. Schiavone, *Phys. Rev. B*, **7**, 4485 (1973).

62. J. E. Jaffe and A. Zunger, *Phys. Rev. B*, **28**, 5822 (1983).

63. C. H. L. Goodman and R. W. Douglas, *Physica*, **20**, 1107 (1954).

64. I. G. Austin, C. H. L. Goodman, and A. E. Pengelly, *J. Electrochem. Soc.*, 103, 609 (1956).

65. D. Raj, K. Chandra, and S. P. Puri, *J. Phys. Soc. Jpn.* **24**, 39 (1968).

66. K. Sato and T. Teranishi, *Jpn. J. Appl. Phys.*, **19**, 101 (1980).

67. Ph. Sainctavit, J. Petiau, A. M. Flank, J. Ringeissen, and S. Lewonczuk, *Physica B*, **158**, 623 (1989).

68. L. O. Brockway, *Zeit. Krist.*, **89**, 434 (1934).

69. R. Nitsche, D. F. Sargent, and P. Wild, *J. Cryst. Growth*, **1**, 52 (1967).

70. W. Schäfer and R. Nitsche, *Mat. Res. Bull.*, **9**, 645 (1974).

71. J. Allemand and M. Wintenberger, *Bull. Sci. Fr. Min. Crystallogr.*, **93**, 14 (1970).

72. P. Imbert, F. Varret, and M. Wintenberger, *J. Phys. Chem. Solids*, **34**, 1675 (1973).

73. J. Ackerman, S. Soled, A. Wold, and E. Kostiner, *J. Solid State Chem.*, **19**, 75 (1976).

74. D. M. Schleich and A. Wold, *Mat. Res. Bull.*, **12**, 111 (1977).

75. Y. Shapira, E. J. McNiff, E. D. Honig, K. Dwight, and A. Wold, *Phys. Rev. B*, **37**(1), 411 (1988).

76. R. Ottenburgs and H. Goethals, *Bull. Soc. Fr. Min. Crystallogr.*, **95**, 458 (1972).

77. K. Doverspike, R. Kershaw, K. Dwight, and A. Wold, *Mat. Res. Bull.*, **23**, 959 (1988).

11

Ternary Transition Metal Chalcogenides AB$_2$X$_4$

Ternary chalcogenides of the type AB$_2$X$_4$ (A, B = transition metals and X = S, Se, Te) generally adopt either the cubic spinel structure or defect structures related to NiAs. It will be shown that both the metal ions and the polarizability of the anion determine which of these structures is favored. In the cubic close-packed anion array of the thiospinel lattice (Fig. 11.1), the metal ions occupy both octahedral and tetrahedral interstices. However, in the nearly hexagonal close-packed array of the NiAs type structures, the metal ions occupy only octahedral interstices. Octahedral vs tetrahedral preference energies of the metal ions that occupy the interstices appear to influence the choice of structure. Spinel formation requires one kind of ion, e.g., Cu(II) $3d^9$, which is stabilized on tetrahedral sites, and a second ion, e.g., Ti(III)($3d^1$), which is stabilized on octahedral sites. Hence, an ion with a strong preference for tetrahedral sites, e.g., Cu(II), Zn(II), might result in formation of a spinel. However, when both metal ions are stabilized on octahedral sites, the NiAs-type structure can form. Furthermore, increasing the anion polarizability in the sequence S<Se<Te increases octahedral site stabilization, which in turn favors the NiAs-type structure over the spinel. In this presentation, the preparation and properties of the thiospinels will first be discussed, followed by a treatment of the ternary NiAs-type structures.

A. Thiospinels

Many oxide spinels have been prepared and studied including those of the metals, V, Cr, Mn, Fe, Co, Ni, Cu. Rogers et al. (1) showed that for the normal vanadium spinels, if the A-site cations are made smaller, the unit cell shrinks and the V(III)–V(III) distance decreases. This corresponded to an increase in the overlap of the V(III) ions. It was then possible to define a critical distance for overlap, R_c, below which the d-electrons of the transition metal are collective rather than localized. This concept of a critical distance was originally developed by Mott (2) and

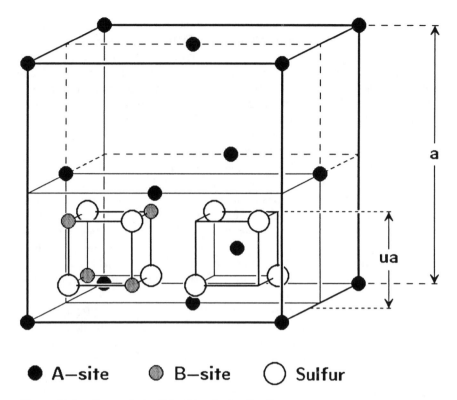

● **A–site**   ◉ **B–site**   ○ **Sulfur**

Figure 11.1.   Two octants of the thiospinel unit cell.

Goodenough (3–5) to explain certain unusual magnetic and electrical properties of transition metal oxides where the metal ions have a $3d^n$ configuration with $n \leq 3$.

Unfortunately, fewer thiospinels exist than oxide spinels. There is only one titanium and one vanadium thiospinel, namely, $CuTi_2S_4$ and CuV_2S_4. A number of thiospinels of the type $A(II)Cr_2S_4$ have been reported in the literature (6–13). The thiochromites are all probably normal, i.e., only divalent ions on tetrahedral sites. This is a result of the octahedral site preference of Cr(III) and has been definitely established for $FeCr_2S_4$ (11) and $MnCr_2S_4$ (12) by neutron diffraction studies. Hahn (14) did show by X-ray diffraction studies that $ZnCr_2S_4$ is normal.

The Cr(III)–Cr(III) interactions in $MnCr_2S_4$ and $CoCr_2S_4$ are ferromagnetic (6,12), which suggests that $R > R_c$ for the following reasons: direct Cr(III)–Cr(III) interactions are antiferromagnetic, whereas the indirect 90° Cr(III)–anion–Cr(III) interactions are ferromagnetic. In oxides, the net interaction is antiferromagnetic, which points to a strong direct Cr(III)–Cr(III) interaction. The ferromagnetic Cr(III)–Cr(III) interactions in $MnCr_2S_4$ and $CoCr_2S_4$ indicate that the

90° indirect interactions are much stronger, relative to the direct interactions, in sulfides than in oxides. This implies that for Cr(III), R_c (sulfides) < 3.51 Å.

In addition to the thiochromites, several thiocobaltites have been prepared. Lotgering (6) has shown that, based on ionic S^{2-} close packing, the cell edge of a thiospinel should be ≥ 9.85 Å. The cell edges of the thiochromites are consistent with this value, from which he concluded that the compounds are essentially ionic. However, the cell edges of the thiocobaltites are about 9.4 Å, which is considerably smaller than that anticipated for purely ionic structures. Since a smaller sulfur–sulfur distance than that necessary for S^{2-} contact is an indication of appreciable covalency, the simple ionic model does not hold for the thiocobaltites. Lotgering has demonstrated that from the cell sizes alone, it is possible to conclude that there must be considerably more covalent character in the cobalt–sulfur bonds. Low susceptibility values indicate that the atomic moments are considerably lower than the spin-only values, and the Curie constants agree with those calculated for covalent structures. For cobalt oxide spinels, it has been shown (15) that the cobalt is in the low-spin state, and appreciable covalency in the thiocobaltites is consistent with low-spin state Co(III) being present in these compounds.

Goodenough (16,17) proposed a band model, and the reader is referred to his papers for the details of this model. Briefly, the e_g levels of the B-site chromium ions in the thiochromites are broadened into a band by covalent mixing with the anion. The magnetic properties and cell sizes of the thiochromites would be the same as in Lotgering's model, since both assume that the t_{2g} electrons on the B-site are localized. For the thiocobaltites, the Goodenough model indicates low spin Co(III) and therefore empty σ^* bands in this case also. Any metallic conductivity is due to partially filled σ^* bands associated with the A-site transition metal. The σ^* bands for B-site cations have e_g symmetry; those for A-site cations have t_{2g} symmetry. The nature of any conductivity will be affected by the A-site ion.

With these considerations in mind, let us examine two one-electron energy diagrams to illustrate different cases considered by Goodenough. Figure 11.2 is the diagram for Co_3S_4 where it is assumed that $R > R_c$. There are no exchange splittings, i.e., all the localized electrons are paired in Co(II) and low spin Co(III). The number in brackets refer to the total degeneracy of a level. These are obtained by multiplying the number of orbitals per atom contributing to a level or band by the number of atoms per molecule. If no exchange splitting occurs, i.e., no localized, unpaired electrons, these numbers must be doubled to include spin degeneracy. Co_3S_4 is metallic. Low spin Co(III) has a high preference for octahedral site occupancy and Co(II) occupies tetrahedral sites. These are formal valencies only since Co_3S_4 is metallic. The t_{2g} and e_g levels are completely filled, but only three 3d electrons from the two sites occupy the antibonding levels. Hence, the observed metallic behavior of Co_3S_4 is consistent with this model.

$CoCr_2S_4$ is illustrated in Fig. 11.3 and is more complex because the exchange splittings of both A- and B-site ions must be considered. As discussed previously,

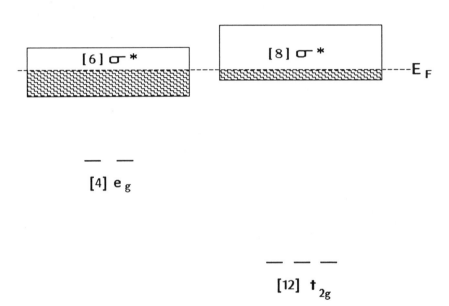

Figure 11.2. Schematic one-electron, one-molecule energy diagram for Co_3S_4.

$R > R_c$ for Cr(III) in the sulfur lattice. The Cr(III) levels are split by an intraatomic exchange energy involving d-electrons on the same ion. The Co(II) levels are split by interatomic antiferromagnetic superexchange, i.e., indirect interactions through an anion intermediary. Indirect interactions of the A–anion–B are always antiferromagnetic according to Anderson (18) and this was the principal interaction considered by Néel (19) in his well-known coupling scheme used to explain the magnetic properties of many oxide spinels. Although other types of interaction can become important in certain cases, i.e., direct B–B and indirect B–anion–B, those of the A–A and A–anion–A are usually not considered. This is because the A-site ions are well separated in the spinel structure. A–sulfur–B interactions will be stronger than A–oxygen–B interactions since the overlap of the A- and B- ions with the anion is greater for sulfur than for oxygen. Since this interaction is antiferromagnetic, A-site metal–ion electrons that are antiparallel to B-site metal-ion electrons are more stable, and hence are lower in energy than those A-electrons that are parallel to B-electrons. This is equivalent to an antiferromagnetic interatomic exchange splitting, that splits an A-site level into two spin states. Hence, a more stable level with spins antiparallel to those of the B-ions and a

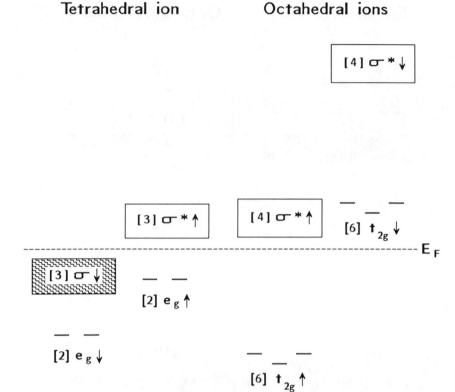

Figure 11.3. Schematic one-electron, one-molecule energy level diagram for $CoCr_2S_4$.

less stable level with spins parallel are formed. For $CoCr_2S_4$, semiconductor behavior is predicted, and Bouchard et al. (13) measured a room temperature resistivity on well-sintered bars of 5×10^3 Ω-cm and a high-temperature activation energy of 0.32(1) eV. If this scheme is applied to the remainder of the thiospinels, the predictions made as to conductivity are in all cases consistent with the observed electrical properties.

The same model can be applied to the selenium spinels. For this series of compounds, σ^* bands form more readily because selenium is more polarizable than sulfur. Therefore, it is likely that these bands are broader than they were in the sulfides. The distance between localized energy levels and the next available band will be decreased and thus the energy necessary for conduction is reduced. It was found by Bouchard (20) that the activation energy for $ZnCr_2Se_4$ is only one-fifth that for $ZnCr_2S_4$.

An interesting group of thiospinels are those containing copper on the A-site. Lotgering and Van Stapele (21) published a model that supports the presence of

$Cu(I)(3d^{10})$ on A-sites in spinels of the type CuM_2X_4 and $CuMM'X_4$, e.g., $Cu(I)[Cr(III)Cr(IV)]S_4$, $Cu(I)[Cr(III)Cr(IV)]Se_4$, and $Cu(I)[Cr(III)Ti(IV)]S_4$. Robbins et al. (22) and Colominas (23) carried out neutron diffraction studies and electrical measurements on $CuCr_2Se_4$ and indicated that their results are consistent with the configuration $Cu(II)[Cr(III)Cr(III)]Se_4$ with the Cu(II) orbitals delocalized to form a partially filled band. Goodenough (24) interpreted the ferromagnetism and metallic behavior of $CuCr_2X_4$ (X=S, Se, Te) by suggesting that since Cu(II) has a $3d^9$ configuration, there is one hole in the $3d$ shell. These holes are delocalized and occupy states of a narrow d-band formed from Cu–X–Cu indirect interactions. These holes give rise to metallic conduction. In addition, at low temperatures, there is a ferromagnetic alignment of $Cr(III)(3d^3)$ electrons giving a moment of $6\mu_B$/mol. There is also an antiparallel spin density due to the holes in the narrow d-band leading to a magnetic moment of $-1\mu_B$/mol. The net moment of $5\mu_B$/mol is consistent with the observed moment. Lotgering and Van Stapele (25,26) modified their model so that $CuCr_2X_4$ contained diamagnetic $Cu(I)(3d^{10})$, and both Cr(III) ions $(3d^3)$ were aligned ferromagnetically. There was one hole present in the valence band, the top of that was made up of chalcogen, nonbonding p-orbitals These holes were responsible for the observed metallic behavior, as well as for the ferromagnetic exchange between Cr(III) ions. Hollander et al. (27) concluded from X-ray photoelectron spectroscopy that the copper present in $CuCr_2Se_4$ is monovalent. Hence, the modified model of Lotgering and Van Stapele is probably correct.

B. The Monoclinic Defect Nickel Arsenide Structure

The defect structures related to NiAs can be classified into three possible types for the stoichiometry AB_2X_4 (28). The arrangement of the metal ions and vacancies found in each of these structures can be seen with the aid of Fig. 11.4. An idealized representation of the close-packed layers of NiAs is shown in Fig. 11.4a. The removal of one-half of the cations by removing alternate cation planes leads to the $Cd(OH)_2$ structure shown in Fig. 11.4c. If only one-fourth of the cations are removed, the resulting structure is M_3X_4 (or AB_2X_4) as shown in Fig. 11.4b. There are three ways in which the cations can be removed and each would result in a different structure type. The completely random removal of one-fourth of the cations results in their being distributed randomly throughout every metal layer. The hexagonal symmetry of NiAs is preserved and the X-ray diffraction pattern shows no 00ℓ reflections for $\ell = 2n + 1$ (n = integer). The space group is $P6/mmc$ and has never been observed for the stoichiometry AB_2X_4. The random removal of one-fourth of the cations from alternate metal layers confines the random distribution of the vacancies to every second metal layer. The symmetry is trigonal (space group $P\bar{3}m1$) and the diffraction pattern is that of a $Cd(OH)_2$-type structure. All of the 00ℓ reflections are present, and this is referred to as the

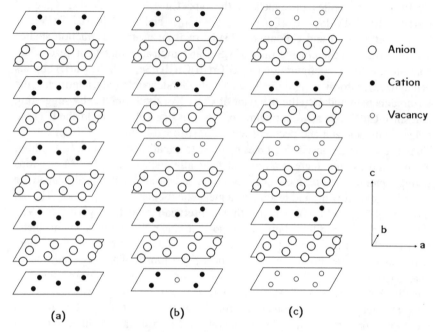

(a)   **(b)**   **(c)**

Figure 11.4.   Idealized representation of the close-packed layers of (a) NiAs, (b) Cr_3S_4, and (c) $Cd(OH)_2$.

trigonal defect NiAs structure. The last way is to remove one-fourth of the cations from alternate metal layers, but in an ordered manner. The vacancies produced are confined to every second metal layer and are ordered within these layers. A distortion to monoclinic symmetry results, as well as a doubling of the unit cell along the c-direction. The diffraction pattern resembles that of Cr_3S_4 (space group $I/2m$) with all reflections satisfying the condition $h + k + \ell = 2n$ (n = integer). The ordered arrangement of Cr_3S_4 is represented in Fig. 11.4b. This structure is referred to as the monoclinic defect NiAs, or Cr_3S_4-type structure.

The metal packing sequence of the idealized monoclinic defect NiAs structure is represented in Fig. 11.5 for the stoichiometry AB_2X_4. It can be seen that filled B^{3+} layers alternate with layers containing the A^{2+} cations and vacancies. An ordering exists between the A^{2+} cations and these vacancies.

The preparation and structural properties of more than 50 chalcogenides of the type AB_2X_4 (A = V, Cr, Fe, Co, Ni; B = Ti, V, Cr; X = S, Se, Te) with defect NiAs structure have been reported (28–39). Most of the compounds were found to have the monoclinic defect NiAs structure, although several were reported to be trigonal.

In considering the band structure of compounds with the defect NiAs structure, only octahedral site ions are involved. Unlike the spinel, all metal ions are on

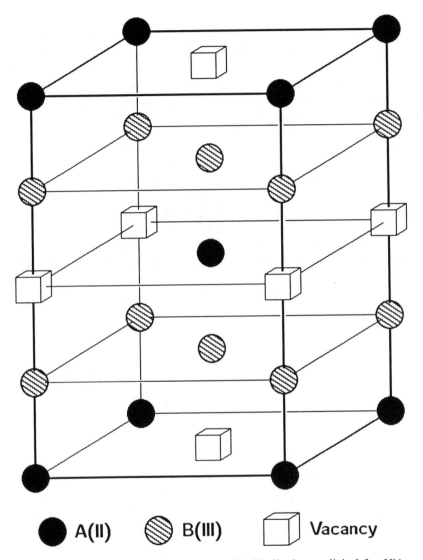

● A(II)   **◉** B(III)   **▢** Vacancy

Figure 11.5. The metal packing sequence of the idealized monoclinic defect NiAs structure for the stoichiometry AB_2X_4.

octahedral sites and, therefore, the t_{2g} levels are lowest in energy. The metallic behavior for the MTi_2X_4 compounds can be ascribed in some measure to partially filled bands formed as a result of Ti-Ti overlap. This is also consistent with the observation that NiV_2S_4 is more metallic than $NiCr_2S_4$. It has been shown by Morin (40) that for transition metal compounds, those containing elements at the beginning of the series have electrical properties consistent with greater *d*-orbital overlap between metal ions. For an isostructural series of compounds, the *d*-orbitals extend further out in space for the earlier transition metals because of the reduced nuclear charge (compared to later transition metals). Therefore, the overlap of Ti(III) is greater than V(III), which in turn is greater than Cr(III). Goodenough (41,42) indicated that the observed transport properties can also be explained by the formation of σ^* bands composed of cation and anion wave functions. According to this model, metallic or semiconducting properties depend on the manner in which these bands are occupied. One-electron energy level diagrams for the MCr_2S_4 compounds were discussed by Holt et al. (43). In these compounds there is no ambiguity possible for the spin states of the various metal species, except $Cr(II)(3d^4)$. It was assumed that for Cr(II), the exchange energy is large enough to result in the high spin state. For metal ions with unpaired *d*-electrons, an intraatomic exchange splitting exists that removes the spin degeneracy. In a one-electron energy diagram, this results in the formation of two different spin states, which can be labeled $t_{2g}(\uparrow)$ and $t_{2g}(\downarrow)$. These can be localized states or narrow bands (see Figs. 11.6–11.8). One level has its electron

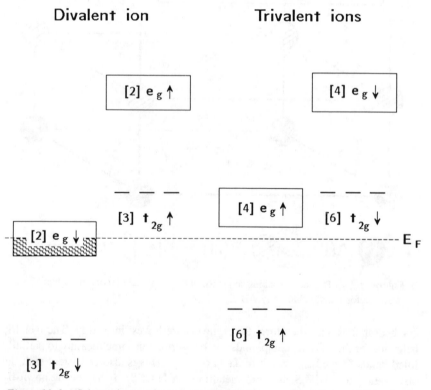

Figure 11.6. Schematic one-electron, one-molecule energy level diagram of the *d*-state manifold for Cr_3S_4.

spins antiparallel to the other. Similarly the e_g levels (narrow σ^* bands) are split into $e_g(\uparrow)$ and $e_g(\downarrow)$ states. Figures 11.6–11.8 represent conventional one-electron, one-molecule energy level diagrams of the d states for Cr_3S_4, $NiCr_2S_4$, and VCr_2S_4. The broad valence and conduction bands formed from s and p states are omitted from the figures for simplicity.

It can be seen from Fig. 11.6 that the $e_g(\uparrow)$ level or $\sigma^*(\downarrow)$ narrow band is half-filled, and therefore Cr_3S_4 should be a metal. The magnetic properties have been reported by Bertaut et al. (44) who found a magnetic unit cell doubled in the a and c directions. The chromium ions are aligned in ferromagnetic sheets parallel to the (101) planes, with only divalent or only trivalent ions in any one plane. These sheets are coupled antiferromagnetically. Hence in Fig. 11.6, the Cr(II) is assigned an antiparallel spin with respect to the Cr(III) species. For $NiCr_2S_4$ (Fig. 11.7), the divalent nickel $t_{2g}(\uparrow)$ and $t_{2g}(\downarrow)$ levels, as well as the divalent $e_g(\downarrow)$ band, are filled, while the divalent $e_g(\uparrow)$ band is empty. Therefore,

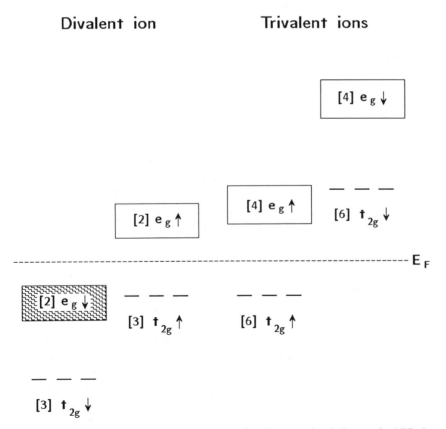

Figure 11.7. Schematic one-electron, one-molecule energy level diagram for $NiCr_2S_4$. Note that the Ni(II) levels are more stable than the Cr(III) levels.

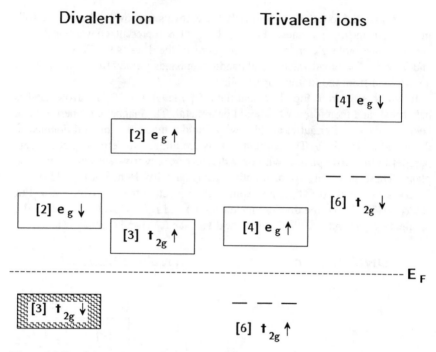

Figure 11.8. Schematic one-electron, one-molecule energy level diagram for VCr_2S_4.

$NiCr_2S_4$ should be a semiconductor, since there is an energy gap between a full band and the next empty available level. For VCr_2S_4 (Fig. 11.8), the divalent vanadium $t_{2g}(\downarrow)$ band is filled and all other levels are empty. Therefore, it should also be a semiconductor. These predictions are consistent with the results reported by Holt et al. (43). It should be noted that for VCr_2S_4, the divalent vanadium t_{2g} levels are represented as bands because of metal–metal overlap. However, they are still split by an exchange energy even though there are no localized vanadium electrons. The splitting in this case is a result of antiferromagnetic interatomic exchange with the localized Cr(III) electrons.

The metallic behavior observed for the MTi_2X_4 compounds (46) can be ascribed in some measure to partially filled bands formed as a result of direct Ti–Ti overlap. The predictions of the Goodenough critical distance model are consistent with this behavior. However, the contribution of the M cations to the conductivity cannot be overlooked. Their participation via σ^* conduction bands formed by cation e_g–anion s,p_σ orbital interaction must be considered. An observed increase in the electrical resistivity of $NiTi_2S_4$ and $NiTi_2Se_4$ (46) accompanying a temperature-dependent order–disorder transition at 67°C indicates that the nickel cations do contribute to the metallic conductivity.

It has also been shown (47) that the defect NiAs structure can also accommodate the second transition series metal, rhodium. The compounds MRh_2X_4 (M = Cr, Co, Ni; X = Se, Te) and Rh_3Te_4 have been prepared and found to have the monoclinic Cr_3S_4-type structure (space group $I2/m$) except for $CoRh_2Te_4$ and $NiRh_2Te_4$ that are trigonal (space group $P\bar{3}m1$). As indicated earlier, the symmetry of these compounds with defect NiAs structure depends on whether the metal ion vacancies are ordered or randomly arranged in alternate metal layers. The metallic behavior observed for these compounds may arise from partially filled σ^* conduction bands formed as a result of interactions between the metal e_g and anion s, p-orbitals.

References

1. D. B. Rogers, R. J. Arnott, A. Wold, and J. B. Goodenough, *J. Phys. Chem. Solids*, **24**, 347 (1963).

2. N. F. Mott, *Proc. Phys. Soc. (London)*, **A62**, 416 (1949).

3. D. Wickham and J. B. Goodenough, *Phys. Rev.*, **115**, 1156 (1959).

4. J. B. Goodenough, *Magnetism and the Chemical Bond*, Interscience Monographs on Chemistry, Inorganic Chemistry Section, Vol. 1, F. A. Cotton (ed.). Interscience, John Wiley, New York, 1963.

5. J. B. Goodenough, *Phys. Rev.*, **117**, 1442 (1960).

6. F. K. Lotgering, *Philips Res. Rep.*, **11**, 218 (1956).

7. F. K. Lotgering, *Philips Res. Rep.*, **11**, 337 (1956).

8. H. Hahn, C. De Lorent, and B. Harder, *Z. Anorgallgem. Chem.*, **283**, 138 (1956).

9. H. Hahn and B. Harder, *Z. Anorgallgem. Chem.*, **288**, 257 (1956).

10. F. K. Lotgering, *Solid State Commun.*, **2**, 55 (1964).

11. G. Shirane and D. E. Cox, *J. Appl. Phys.*, **35**, 954 (1964).

12. N. Menyuk, K. Dwight, and A. Wold, *J. Appl. Phys.*, **36**(3) Part 2, 1088 (1965).

13. R. J. Bouchard, P. A. Russo, and A. Wold, *Inorg. Chem.*, **4**, 685 (1965).

14. H. Hahn, *Z. Anorg. Allegem. Chem.*, **264**, 184 (1951).

15. W. L. Roth, *J. Phys., Chem. Solids*, **25**, 1 (1964).

16. J. B. Goodenough, *Propiétés Thermodynamiques Physiques et Structurales des Derives Semi-Metallique*, Centre National du Science, Paris, 1967, pp.263-922.

17. J. B. Goodenough, *J. Phys. Chem. Solids*, **30**, 261 (1969).

18. P. W. Anderson, *Phys. Rev.*, **115**, 2 (1959).

19. L. Néel, *Ann. Phys. (Paris)*, **3**, 137 (1948).

20. R. Bouchard, Ph.D. Thesis, Brown University, June 1966.

21. F. K. Lotgering and R. P. Van Stapele, *Solid State Commun.*, **5**, 143 (1967).

22. M. R. Robbins, H. W. Lehmann, and J. G. White, *J. Phys. Chem. Solids*, **28**, 897 (1967).

23. C. Colominas, *Phys. Rev.*, **153**(2) 558 (1967).

24. J. B. Goodenough, *Solid State Commun.*, **5**, 577 (1967).

25. F. K. Lotgering and R. P. Van Stapele, *J. Appl. Phys.*, **39**, 417 (1968).

26. R. P. Van Stapele and F. K. Lotgering, *J. Phys. Chem. Solids*, **31**, 1547 (1970).

27. J. C. Th. Hollander, G. Sawatzky, and C. Haas, *Solid State Commun.*, **15**, 747 (1974).

28. F. Jellinek, *Acta Crystallogr.*, **10**, 620 (1957).

29. S. L. Holt, R. J. Bouchard, and A. Wold, *J. Phys. Chem. Solids*, **27**, 755 (1966).

30. R. J. Bouchard and A. Wold, *J. Phys. Chem. Solids*, **27**, 591 (1966).

31. H. Hahn and B. Harder, *Z. Znorg. Allgem. Chem.*, *x 288*, 241 (1956).

32. H. Hahn, B. Harder, and W. Brockmuller, *Z. Anorg. Allgem. Chem.*, **288**, 260 (1956).

33. M. Chevreton and A. Sapet, *Compt. Rend.*, **261**, 928 (1965).

34. M. Chevreton and F. Bertaut, *Compt. Rend.*, **255**, 1275 (1962).

35. G. Bérodias and M. Chevreton, *Compt. Rend.*, **261**, 2202 (1965).

36. F. Grønvold and E. Jacobsen, *Acta Crystallogr.*, **10**, 1440 (1956).

37. M. Chevreton and G. Bérodias, *Compt. Rend.*, **261**, 1251 (1965).

38. H. Haraldsen, F. Grnvold, and T. Hurlen, *Z. Anorg. Chem.*, **283**, 143 (1956).

39. W. Klemm and N. Fratini, *Z. Anorg. Chem.*, **251**, 222 (1943).

40. F. J. Morin, *J. Appl. Phys*, **32**, 2195 (1961)

41. J. B. Goodenough, *Bull. Soc. Chim. France*, **4**, 1200 (1965).

42. A. Ferretti, D. B. Rogers, and J. B. Goodenough, *J. Phys Chem. Solids*, **26**, 2007 (1965).

43. S. L. Holt, R. J. Bouchard, and A. Wold, *J. Phys. Chem. Solids*, **27**, 755 (1966).

44. F. Bertaut, E. F. Roult, G. Leonard, R. Pauthenet, M. Chevreton, and R. Jansen, *J. Appl. Phys.*, **35**(3), Pt.2, 952 (1964).

45. R. H. Plovnick, D. Perloff, M. Vlasse, and A. Wold, *J. Phys. Chem. Solids*, **29**, 1935.

46. R. H. Plovnic and A. Wold, *Inorg. Chem.*, **7**, 2596 (1968).

Problems

P11.1. FeS_2, CoS_2, and NiS_2 crystallize with the pyrite structure, and the transition metal ions have $3d^6$, $3d^7$, and $3d^8$ electronic configurations. However, other sulfides of iron, cobalt, and nickel do not behave according to the usual expectations of their chemistry. Choose three sulfide examples containing Fe, Co, and Ni and discuss
   a. their preparation
   b. crystal structure
   c. unusual electronic structure
   d. magnetic and electronic properties.

P11.2. The dichalcogenide MoS$_2$ is a semiconductor whereas the isostructural compound NbS$_2$ is a superconductor. On the basis of a crystal field model, explain the difference in the conductivities of these two compounds.

P11.3.   a. What is the most common sulfide of ruthenium and what are its structure and electronic properties?

     b. What is the most common sulfide of rhodium and what are its structure and electronic properites?

     c. How many sulfides exist for platinum and what are the coordination of the platinum with respect to sulfur for each sulfide?

     d. CuS contains both copper(I) and copper(II). Write a formula that is consistent with equal quantities of both species. Why does CuO contain only Cu(II) and yet the sulfide have both species?

P11.4. Discuss the structural relationships between the following pairs of substances:

     a. C(diamond)–ZnS

     b. ZnS–CuFeS$_2$

     c. CuFeS$_2$–Cu$_2$ZnGeS$_4$

     d. ZnGeP$_2$–CuGaS$_2$

P11.5. The intercalation of lithium into TiS$_2$ has resulted in a series of compounds that can be used as cathodes for the Li/S battery. Refer to recent papers that describe this particular application for Li$_x$TiS$_2$ and indicate what are the advantages and disadvantages of this material over other known cathode materials for the Li/S battery.

P11.6.   a. In the system CuGa$_{1-x}$Fe$_x$S$_2$, the magnetic behavior of Fe(III) changes with the value of x. What would be the nature of the susceptibility data above and below $x = 0.3$.

     b. For the end member CuFeS$_2$, it is essential that the total metal to sulfur ratio be 2:2. Growth of single crystals by the Bridgman technique results in products that are not stoichiometric. What is the nature of the observed nonstoichiometry and how may this be remedied so as to give crystals with the proper sulfur to metal ratio?

P11.7.   a. Discuss the two crystal modifications of ZnS. Would you expect any differences in the optical band gaps of these two isotypes? Why?

     b. How do the electronic properties of SnS$_2$ differ as a function of metal:sulfur ratio. What would be the differences in the electronic properties of SnS$_2$ grown by vapor transport and chemical vapor transport?

Answers to Selected Problems

Chapter 1

A1.3. See Fig. A1.3.

A1.4. See Fig. A1.4.

A1.5a. As seen in Fig. A1.5, the [110],[101] and [011] directions all lie in the (111) plane, and all are perpendicular to the [111] direction.

A1.5b. The [010] and [001] directions lie in the (100) plane, and both are perpendicular to the [100] direction.

A1.6. See Fig. A1.6.

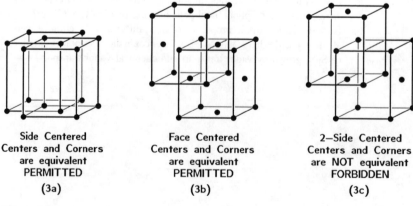

Side Centered	Face Centered	2–Side Centered
Centers and Corners	Centers and Corners	Centers and Corners
are equivalent	are equivalent	are NOT equivalent
PERMITTED	PERMITTED	FORBIDDEN
(3a)	(3b)	(3c)

Figure A1.3

236

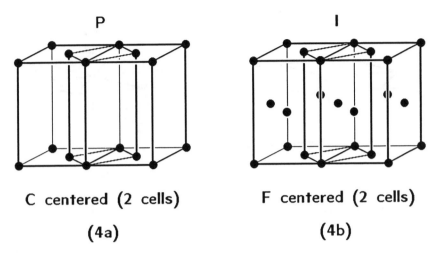

C centered (2 cells)   F centered (2 cells)

(4a)   (4b)

Figure A1.4

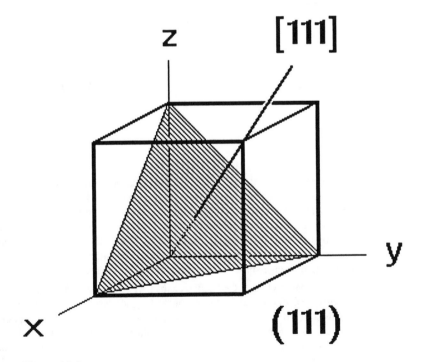

Figure A1.5

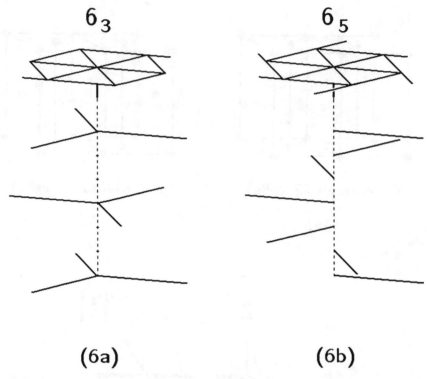

(6a)   (6b)

Figure A1.6

Chapter 2

A2.2.   A mechanical mixture of NaCl and AgCl will give an X-ray diffraction pattern that contains the diffraction peaks from both compounds. However, when the sample is heated to produce a homogeneous solid solution, the pattern will contain peaks expected from a single phase in which 50 at% of the sodium ions has been replaced by the large silver ions in the structure. Hence, the peaks of NaCl will be displaced (shifted) toward smaller 2θ values. The intensity of many of the peaks will also change since the scattering of Ag^+ is greater than Na^+.

A2.4a   and 4b. The ionic radius of Cr^{3+} is larger than Al^{3+} (0.62 vs 0.54). There may be observed a small downward shift of the Al_2O_3 peaks as Cr^{3+} is substituted for Al^{3+}. This shift will continue until maximum solubility is achieved. At that point, there would no longer be a decrease in the 2θ values and diffraction peaks due to pure Cr_2O_3 would be observed. In this system, there is complete solubility and hence the shifting continues throughout the complete substitution of Cr^{3+} for Al^{3+}.

A2.4c   As the concentration of Cr^{3+} decreases, there are fewer $Cr^{3+}-Cr^{3+}$ interactions. The Néel temperature will have a large negative value (strong antiferromagnetic interactions) at high Cr^{3+} levels and as the concentration of Cr^{3+} is decreased, the θ value (Weiss constant) will become less negative because of the weaker antiferromagnetic interactions.

Chapter 3

A3.2. Green NiO $\xrightarrow[O_2]{heat}$ $Ni_{1-x}^{2+} Ni_x^{3+}$ O (black)

The higher conductivity of the black form is due to the presence of mixed valence states of nickel and hence electrons can readily transfer from one state to the other. Pure green NiO is anti-ferromagnetic and above the Néel point shows paramagnetic behavior characteristic of octahedrally coordinated Ni(II).

A3.5. Arsenic atoms can be substituted for germanium atoms but arsenic has five valence electrons and germanium only four. Hence, germanium doped with arsenic is an n-type semiconductor and the conduction is due to an excess of negative charge that lies close to the conduction band and hence can be readily promoted.

Chapter 4

A4.2. VO is metallic and has strong Vt_{2g}–Vt_{2g} overlap. The band that results from this overlap is only partially occupied. Metallic behavior is consistent with temperature independent paramagnetism (Pauli paramagnetism) and high conductivity. NiO shows no overlap of d-wave functions. Ni(II) behaves as a paramagnetic $3d^8$ ion in an octahedral field.

A4.3. Zn[Fe(III) Fe(III)]O$_4$

$$3d^5 \qquad 3d^5$$
$$\uparrow \qquad \downarrow$$

The B-site ions interact antiferromagnetically and hence there is no net moment.

A4.5. $Fe^{3+}[Ni_x^{2+}Fe_{2-x}]O_4 = Fe^{3+}[Fe_{1-x}^{2+}Ni_x^{2+} Fe^{3+}]O_4$

If $x = 1$, then we have a net moment of 2 BM per formula weight.

Chapter 5

A5.1. (a) No. CsCl is immiscible with NaCl because Cs^+ is much larger than Na^+ and cannot be accommodated into the NaCl structure. In CsCl there are 8 Cl^- nearest-neighbors whereas in NaCl there are 6. (b) NaCl and KCl are completely miscible; both crystallize with the rock salt structure.

A5.4. The phase diagram (Fig. A5.4) for CaO and ZrO_2.

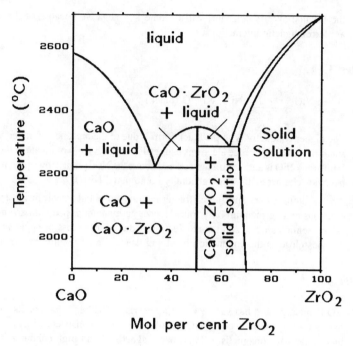

Figure A5.4

Index